Unholy Alliance

The Scientific & Religious Conspiracy

RICK ARONS

Unholy Alliance
The Scientific & Religious Conspiracy
against God and the Jews

Published 2017
ISBN 978-0-9667019-3-7
Copyright © 2017
by Rick Arons

All Rights Reserved

No part of this publication may be translated, reproduced, stored in a retrieval system or transmitted in any form or by any means, electronic, mechanical, photocopying, recording or otherwise, without prior permission in writing from both the publisher and the copyright holder.

Published by BS"D Press, Skokie, Illinois, U.S.A.

For information on this and other publications, contact BS"D Press at BSDPress@aol.com

Contents

Introduction: A Mind is a Terrible Thing to Waste vii

I	**The Philosophical Quest** 1
	The Teleological Argument or "Fine Tuning" 6
	The Science of Fine Tuning 9

II	**The Reasoned Attack on Religion** 17
	The Opium of the People? 17
	The Unrelenting Attack on Institutional Religion 19

III	**The Response of Scripture: The Chosen Concept** 26
	The Manipulative Power of Alternative Facts 26
	Why all the fuss about a "Chosen" people? 28
	Being Chosen 30
	The Modern Development of the Exceptionalist Idea 31
	Religious Exceptionalism 33
	Contemporary Evidence of Jewish Exceptionalism? 34

IV	**Who are the Real Chosen People?** 37
	A Brief History of Religion 37
	The Nature of Christian Exceptionalism 38
	The Limitations of Christian Exceptionalism 42
	Islamic Exceptionalism 48
	The Campaign against Jewish Exceptionalism 56
	Alternative Facts and Christian Suppression 57
	Alternative Facts and Muslim Oppression 62

V	**Mesora and the Limitations of Jewish Leadership** 66
	What hath History Wrought! 66
	Survival Mode 69

	The Mesora as a Weapon	72
	A Modern Challenge to the Mesora	79
VI	**The Biblical Chosen Promise**	**91**
	Defining Exceptionalism	91
	Scientific and Religious Paradigms	93
	History's Five Jewish Paradigms	96
VII	**The Elements of Jewish Exceptionalism**	**101**
	1. Scripture, its Authenticity and Historical Accuracy	101
	Biblical Authorship	102
	Historicity of the Bible	107
	Archaeology and the Historical Record	110
	Finding Abraham in History	111
	Correcting a Corrupted Chronology	115
	The Case for Joseph and Egypt's Old Kingdom	117
	Placing the Exodus Where it Belongs	119
	Evidence of the Jews in Ancient Egypt	123
	The Jews as Slaves in Egypt	127
	The Convergence of Biblical and Conventional Chronology	129
	Who were the Hyksos?	131
	Chronological Agreement: Exodus & Conquest	134
	In the Times of Biblical Kings	138
	2. The Mother Tongue	141
	"Proto-Hebrew"	144
	The Hebrew Language as an Information System	147
	The Personal Connection	149
	The Natural Science of the Hebrew Language	152
	And about that Pregnancy and Birth . . .	155
	3. The Kabbalistic and Scientific Tradition	157
	The Scientific and Spiritual Fundamental of Unity	159
	The Unity of God and the Unity of Nature	161
	The Vocabularies of Jewish Mysticism and Science	166
	The Nexus of the Physical and Spiritual	173
	Hidden Dimensions	175
	How Science Views Creation: An Introduction	178
	A Theory of Everything? Determinism, Reductionism, Free Will and Quantum Cosmology	180

    Science and Biblical Creation: A Fairy Tale?........................184
    In the Beginning...186
    Something from Nothing..187
    Biblical and Scientific Agreement on the Age of the
    Universe..189
    The Idea of Creation Becomes Creation..............................195
    Scientific Conclusions about Creation.................................197
    Information and the Mystical Tradition 202
    God, Information, Entropy and the Holographic Principle..208
    Circumcision, Entropy and Jewish Exceptionalism 214
    Where do Consciousness, Mind & Soul Fit in?....................219
    Quantum Consciousness 224
    The Music of the Universal Soul... 228
    Panpsychism and its Believers... 230
    Sidebar: Stapp's Theory of Consciousness Brings Quantum
    Theories Together..232
    A Single, Immaterial Consciousness that Controls Reality ..233
    Free Will..235
    Tying Soul, Mind and Consciousness Together 243
    Evil and Providence: Barry's Story......................................247
    Mind over Matter: How We Routinely Manipulate the
    Information that Determines Outcomes................................252
    The Power of Group Think..257
    Linking Free Will, Providence and the Problem of Evil...... 260
  4. The Messianic Idea 265
  5. The Land 272
    Anti-Semitism and "The Jewish Problem" 276

VIII The Future of Jewish Exceptionalism 281

Postscript.. 287

Unholy Alliance: Sources 291

Introduction: A Mind is a Terrible Thing to Waste

In 1932, 11 year-old Ralph Alpher was attending the local religious school and simultaneously cultivating an active interest in theoretical physics. Raised in an orthodox Jewish home, Ralph regularly challenged his rabbi on the biblical recounting of creation. The two often clashed and rarely had a productive discussion. Worse still, Ralph came away from these arguments with little of a positive nature about his faith. "It got hot and heavy," Alpher recalled. "I'd bring in quotations with references and he didn't want to have any part of it. Finally he (the rabbi) said, 'you gotta go through the bar mitzvah to honor your father. And, after that I don't care what the hell you do!'"[1]

Needless to say, Ralph wasn't a regular at the synagogue thereafter. In 1937, at the age of 16, Alpher was offered a full scholarship to M.I.T. The scholarship was withdrawn when the school learned he was Jewish.

Five years later, Alpher's marriage was officiated by a reform rabbi who brought his poodle to the wedding ceremony. Alpher's orthodox parents did not attend.

In 1948 Alpher wrote a dissertation that applied mathematics to the idea that the universe began in a super-hot "Big Bang" 14 billion years ago. Sixteen years later two Bell Laboratories radio astronomers, Penzias and Wilson, accidentally discovered what Alpher had earlier theorized. They won the Nobel Prize. Alpher got nothing.

---

[1] *Union College Magazine*, Winter, 2000, "The Last Big Bang Man Left," Joseph D'Agnese.

A college instructor in his later years, Alpher is still barely acknowledged for one of the most important discoveries in history. His health suffered and his life was in some ways a bitter one.

As Ralph Alpher discovered despite claims to the contrary, neither religion nor science has all the answers. And perhaps more discouraging is the knowledge that those we identify as our religious and scientific icons have an embarrassingly low tolerance for ambiguity. Behavior or thought that is "outside the box" in either field is routinely dismissed for its irreverence. Our lives are based on assertions that we accept as fact only to later learn of their inadequacies. Neither religion nor science suffers changing intellectual paradigms well.

Ralph Alpher's youthful challenge to both religious and scientific orthodoxy ran headlong into a mutual fear of intellectual co-dependence that has fortified the wall between natural science and scripture since Galileo. This barrier has emboldened the proponents of each and, not incidentally, stolen the biblical birthright of a seemingly insignificant, yet apparently indestructible people. Jewish survival must surprise even the most self-assured, scientific atheist. And, yet, this counterintuitive, historical reality is dismissed by the skeptic -- even in the face of both archeological and scientific revelations that showcase a shrinking boundary between what has traditionally been defined as faith and that which is known today as science.

Like Ralph Alpher, men and women of science have become convinced that humans created God and religion in order to comfort mankind in its quest to answer life's existential questions – questions ordinarily left to philosophy and faith. History demonstrates that the limitations of religious and scientific inquiry have allowed very little integration of thought. Instead, proponents of each compete to dismantle the biblical narrative and discredit its modern applicability. This de facto conspiracy has disrupted the divine plan and imposed a terrible cost on mankind. The much sought after "theory of everything" will neither be found in pure science nor blind faith.

For centuries most were convinced that the answers to our existential questions were best found in philosophy. Yet today even the most carefully constructed proofs of the existence of a supernatural creator are denigrated by both the religious and scientific communities.

If there is a God -- if there is any truth to either philosophical or religious reasoning -- then there must be a fundamental commonality between the philosophical, religious and scientific assessments of existence. The evidence is now abundant. By putting aside intellectual predispositions, explanations for many of the most troubling and fanciful religious beliefs become apparent. The Jewish biblical narrative is the source material for each scriptural, monotheistic fundamental, however, subsequent institutional religious development has diminished the value of that source. We can now take a practical and reasoned approach to what has historically been dismissed as nothing more than faith.

I

THE PHILOSOPHICAL QUEST

THE HISTORY OF WHAT WE BELIEVE AND WHY

Philosophers have addressed existential issues and the presence of a Creator for millennia. Both religious and academic proofs utilize the rules of logic to reach conclusions that we generally accept but with which we have grown increasingly uncomfortable in the age of science. Yet, even without an empirical or evidentiary justification, the overwhelming majority of us continue to believe in an unseen God/Creator – typically for very personal reasons.

In a national telephone survey of over 3000 adults in the United States in 2006 and 2008 it was found that 28% see God as someone engaged in history and meting out harsh punishment to those who do not follow him. Twenty two percent believe in a benevolent God who is engaged in our world and loves and supports us. Twenty one percent believe in a critical God who watches this world but delivers justice in the next world.[2] Twenty four percent believe in the distant God, who

---

[2] A survey by Kurt Gray of the University of Maryland and Annie Knickman and Dan Wegner of Harvard investigated how people perceived those in a persistent vegetative state (*The Economist*, 8/20/2011, p. 73). "Irreligious participants gave the buried corpse about the same mental ratings as the vegetative patient. Religious participants, however, continued to ascribe less mind to the irretrievable unconscious David than they did to his buried corpse. That those who believe in an afterlife ascribe mental acuity to the dead is hardly surprising. That those who do not are inclined to do so unless heavily prompted not to is curious indeed."

created things, but left us alone. This is the dominant view of a majority of Jews, Buddhists and Hindus.[3]

As a part of *The Explaining Religion Project* (begun in 2007), Nicolas Baumard (Oxford, and later the University of Pennsylvania) concluded from early testing that people have a tendency to look for "just rewards" based on their behavior. When he teamed with Ryan McKay (University of London) and Pierrick Bourrat (University of Sydney) to check on the idea that "God is watching," they found that people tended to be more careful about their behavior in the presence of this suggestion. They also found that when going over the *World Values Survey* (which asks respondents from 87 countries about many things including religious beliefs) that people whose religions include an omniscient, judgmental God regard transgressions more harshly than those without such a belief, such as the adherents of many Far Eastern faiths.[4]

Ancient philosophers looked for a "first cause" that could be credited with generating nature and the cosmos as we know them. By assigning God that role, the characteristics that must describe Him could be deduced logically. The reasoning went that if a being is perfect, then that being must also be active.[5] A perfect being is both simple and unchanging and since it is active, it must have something to act upon – something other than itself. And this is how the Bible defines God – the Creator of the universe.

According to Einstein, space and time are attributes of matter. A definition of the God of philosophy takes into account that He cannot be contained by physical dimensions because He is not material. His place is outside of creation, outside of the limitations of our reality. Religious "philosophers" even deduced a Hebrew name for God that reflects this: *makom* (place).[6] The Midrash (extra-biblical commentary) explains that the world is *not* His place. And since He is outside of our

---

[3] "America's Four Gods: What We Say About God – And What That Says About Us," Paul Froese and Christopher Bader, Baylor University sociologists, 2010

[4] *The Economist*, April 23, 2011, pp. 84-85

[5] *Leibniz, Monadology, Theodicy: Essays on the Goodness of God, the Freedom on Man and the Origin of Evil*, translated by E.M. Huggard, La Salle, IL, Open Court, 1985; Arey Kaplan, *Intercom*, February, 1972, p. 125, "The God of Israel"

[6] Midrash Breishit Rabbah 68:10

dimensions, including time, He does not change since change and time are codependent. Consequently God does not change his mind.

One thousand years ago, Christian theologians aggressively sought religious proofs for the existence of God and created the foundation for modern philosophy. The Benedictine monk, Anselm proposed that the very acknowledgement of the idea of God is tantamount to His reality. He concluded that it was possible for the mind to know God by reason alone. Immanuel Kant later termed this the "ontological" argument. Anselm's life coincided with the rediscovery of Aristotelian thought and logic. Applying the Greek dialectic to the "problem of God" left Christian thinkers with an understanding that all of creation was in some way a unity over which God presided. This "scholastic" approach brought a new understanding of the relationships between things, the mind, logic and language.

Thomas Aquinas (Catholicism) and Maimonides (Judaism) felt the strong pull of Aristotle in putting some personal distance between their "proofs" and that of Anselm. "Yes, granted, that everyone understands that by this name God is signified something than which nothing greater can be thought, nevertheless, it does not therefore follow that he understands that what the name signifies exists actually, but only that it exists mentally. Nor can it be argued that it actually exists, unless it be admitted that there actually exists something than which nothing greater can be thought; and this precisely is not admitted by those who hold that God does not exist."[7]   Aquinas and Maimonides (writing a century before him), understood that God's existence could be demonstrated through his actions in the world. Aquinas revised Anselm's proof to define God's role as what he called a "first mover," or "first cause," a necessity who was the creator of the ultimate standards, and who was the intelligent and purposeful administrator of creation.

The use of logic to somehow "prove" God's existence was controversial from its inception. Led by the 14th century Franciscan monk, William of Ockham, "nominalists" challenged the idea that invisible substances could link real things. For Ockham only material things could be grasped by the human senses. The mind, the world and God

---

[7] *The Proof of God*, p. 88, quoting Thomas Aquinas' *Summa Theologica*

bore no essential connection. Ockham's razor,[8] as the application of this belief to philosophical and scientific issues came to be called, excluded the philosophical and metaphysical as evidence of truth. Explanations could be reduced to simple "things." Human knowledge is the knowledge of discreet things, not immaterial "essences." As the idea of God cannot be reduced to something material, nominalism denies the possibility of proving the existence of an immaterial God at all.

Unlike the God of the nominalist, the God of Descartes[9] oversaw creation with a set of constant, rational laws. "He rejected Aristotle, but he also rejected nominalism, arguing that eternal truths could be known with great certainty, while 'things' – made not of atoms, but indefinite divisions of matter – were ambiguous until given certainty by reason, mathematics and metaphysical principles."[10] Borrowing from Aquinas, Descartes posited that the imperfect human mind could not be the cause of the idea of God. That idea had to have been put there by a perfect source – by God himself. And like Anselm, he was convinced of both the necessity of God and the understanding that the thought of God is inseparable from His existence. Baruch Spinoza reflected Descartes in his "pantheism" seeing God and the universe as one. Gottfried Leibniz bolstered the Ontological Argument by trying to show with mathematical logic that if God is possible, then He exists.

With the Enlightenment of the 18th century came Hume and Kant, each denigrating the ontological proof by offering that existence by itself plays no role in logic. While the logic of the proof of God might be sound, there need not be a reason to accept its premise that there exists a reality greater than any that can be imagined. Kierkegaard believed in faith, but saw no place for reason in that faith because, he believed, faith itself is irrational.

Twentieth century, Nobel Prize winner Bertrand Russell put it this way: "If a word means something, there must be something that it

---

[8] William of Ockham lived in 14th century England. He was a Franciscan friar and expert logician. He deduced that assumptions should not be multiplied without necessity – the most simple explanation is usually the best.

[9] Rene Descartes was a French writer, philosopher and mathematician, called the father of modern philosophy. He also developed the Cartesian coordinate system and introduced analytical geometry.

[10] *The Proof of God*, p. 151

means."[11] Philosophy leaves us still trying to prove that the word God does indeed mean something. In the words of contemporary philosopher Richard Swinburne, ". . . although reason can reach a fairly well-justified conclusion about the existence of God, it can reach only a probable conclusion, not an indubitable one."[12]

The Ontological Argument spawned the Cosmological Argument which suggests that the universe came into existence from nothing. Since the laws of thermodynamics preclude something from nothing, the "first cause" must lie outside of the system itself.

The Teleological Argument (on which more will be said below) proposes that the universe has a particular design – one that is specific to the nurturing and development of humanity. Only the design that we have been given could have resulted in creation's development as we know it.

Finally, the Moral Argument posits that there is a generally authoritative morality to which mankind subscribes. Its origin and unanimity must be Divine. [13]

---

[11] *My Philosophical Development*, p. 49

[12] *The Existence of God*, p. 2

[13] A recent study of babies tells us a great deal about morality and its source. "What do these findings about babies' moral notions tell us about adult morality? Some scholars think that the very existence of an innate moral sense has profound implications. In 1869, Alfred Russel Wallace, who along with Darwin discovered natural selection, wrote that certain human capacities — including 'the higher moral faculties — are richer than what you could expect from a product of biological evolution. He concluded that some sort of godly force must intervene to create these capacities. (Darwin was horrified at this suggestion, writing to Wallace, 'I hope you have not murdered too completely your own and my child.')" "A growing body of evidence, though, suggests that humans do have a rudimentary moral sense from the very start of life. With the help of well-designed experiments, you can see glimmers of moral thought, moral judgment and moral feeling even in the first year of life. Some sense of good and evil seems to be bred in the bone."

"One lesson from the study of artificial intelligence (and from cognitive science more generally) is that an empty head learns nothing: a system that is capable of rapidly absorbing information needs to have some prewired understanding of what to pay attention to and what generalizations to make. Babies might start off smart, then, because it enables them to get smarter." "We found that, given a choice, infants prefer a helpful character to a neutral one; and prefer a neutral character to one who hinders. This finding indicates that both inclinations are at work — babies are drawn to the nice guy and repelled by the mean guy. Again, these results were not subtle; babies almost always showed this pattern of response." "Regardless of how smart we

The challenges faced by the philosopher are many. In the end, however, "We shall see that the scientifically inexplicable, the odd and the big, form the normal starting points for arguments to the existence of God. Cosmological and most teleological arguments argue from phenomena allegedly too big for science to explain; whereas most other arguments argue from phenomena allegedly too odd for science to explain. The arguments need to show also that there is no personal explanation in terms of the action of an embodied agent. This done, what has been shown is that an explanation in terms of a very powerful non-embodied agent is the only possible explanation of the phenomena. It then follows that either theism or something like it is true, or that the phenomena are just brute inexplicable facts, the stopping point of explanation."[14]

Anselm's Ontological Argument defined God "as that which nothing greater can be conceived," and argued that this Being could exist in the mind. Modern philosophers reason that if the greatest possible being exists in the mind, it must also exist in reality. If it only exists in the mind then a greater being is possible, existing in both the mind and reality.

THE TELEOLOGICAL ARGUMENT OR "FINE TUNING"

True believers in intellectualizing their philosophical conclusions have largely failed to convert the skeptics. As a result, they increasingly rely on what is known as the Teleological Argument or the Argument from Design. "Teleological" comes from the Greek word meaning purpose (telos, from the Hebrew toelet) indicating that creation was carried out by an intelligent being in order to accomplish a purpose. It is also known as the Argument from Fine-Tuning. This idea was popularized

---

are, if we didn't start with this basic apparatus, we would be nothing more than amoral agents, ruthlessly driven to pursue our self-interest. But our capacities as babies are sharply limited. It is the insights of rational individuals that make a truly universal and unselfish morality something that our species can aspire to." *New York Times Magazine*, May 5, 2010, "The Moral Life of Babies," Paul Bloom. (Infant Cognition Center, Yale University)

[14] *The Existence of God*, p. 75

by 19th century English clergyman and philosopher William Paley. He compared the universe to a watch with hundreds of parts whose integration and order allow it to achieve a very specific goal. Contemporary thinkers reason that we humans live in a remote portion of a nearly infinite universe that was in existence for billions of years before we came along. We are just one of myriad species on earth, a planet with a fossil record that testifies to the fact that 99 percent of all species that have ever lived are now extinct. There must be a reason why we are not among them.

"Scientists, historians, and detectives observe data and proceed thence to some theory about what best explains the occurrence of these data. We can analyze the criteria which they use in reaching a conclusion that a certain theory is better supported by the data than a different theory – that is, is more likely, on the basis of those data to be true. Using those same criteria, we find that the view that there is a God explains everything we observe, not just some narrow range of data. It explains the fact that there is a universe at all, that scientific laws operate within it, that it contains conscious animals and humans with very complex intricately organized bodies, that we have abundant opportunities for developing ourselves and the world, as well as the more particular data that humans report miracles and have religious experiences. In so far as scientific causes and laws explain some of these things (and in part they do), these very causes and laws need explaining, and God's action explains them. The very same criteria which scientists use to reach their own theories lead us to move beyond those theories to a creator God who sustains everything in existence."[15]

A 2005 *Newsweek Magazine* poll found that 80% of the American public believes that God created the universe. Nevertheless reasoned rebuttals leave most believers with only intuition and emotion as a viable defense. The Teleological Argument combats these humanistic firewalls.

A bastardized version of the Teleological Argument has been politicized and employed in the public sphere. "Intelligent Design" is offered as a seemingly reasoned, scientific surrogate for philosophy and faith. Supporters of this sometimes intellectually dishonest attempt to gain

---
[15] *Is There a God*, p. 2

and retain a foothold in public education failed in their quest to give Intelligent Design (and religion) an umbrella of technical legitimacy and elevate it to the level of science.

In December, 2005, The U.S. District Court for the Middle of Pennsylvania heard the case of Tammy Kitzmiller vs. the Dover Area School District. It was ruled that the Dover mandate (which required a statement questioning the factual accuracy of the theory of evolution be read in class) was unconstitutional. Intelligent design was then barred from instruction in the Dover school district's public school science classrooms. The District Court made clear that Intelligent Design is not to be considered a science but rather the teaching of religious creationism, which violates the establishment clause of the U.S. constitution. This dismissal of Intelligent Design is in part a result of its association with "young earth" creationism, a roundly discredited reading of the Christian Bible that embraces a 6000 year old earth. Fine Tuning or the Teleological Argument is actually in complete agreement with evolution and other notions that are rejected by many on the religious right.

Arguments for fine tuning are associated with the idea that proof of a creator rests largely on the infinitesimally tiny possibility that existence could have developed in the way that it did by chance, favoring life and its incredible complexity and sophistication.

While evolutionary biology dismisses the Teleological Argument, it has its own design hypotheses. Simon Conway-Morris, a Cambridge University paleontologist, believes that evolution has trends, two of which are toward complexity and intelligence. He does not believe that things just happen in chemistry. "They happen because of pre-existing causes. Whether it is the molecules of crystalline that are used to build an eye or the hemoglobin that makes blood carry oxygen, the nature of molecules themselves means that evolution is more likely to follow a path determined by their basic structure. Evolution is a mechanism and it works within rules."[16]

Chance is the fallback position of those who deny a designer universe. Chance says that if you have enough monkeys and enough time, the monkeys could produce Hamlet on a typewriter. But Brett Watson[17]

---

[16] *The Economist*, "Unfinished Business," February 7, 2009, p. 74

[17] "The Mathematics of Monkeys and Shakespeare," Brett Watson, 12/13/95,

demonstrates how futile this effort is. He shows that in 17 billion galaxies, each containing 17 billion habitable planets, each with 17 billion monkeys each typing away and producing one line per second for 17 billion years, the chances of the phrase, "To be or not to be, that is the question," not being included in the output is 0.999999999999465759379507 ad nauseum. So it is about 99.999999999995% certain that they would fail. Just one sentence and it is essentially impossible. If even a tiny fragment of Shakespeare's writing is impossible to randomly create, to do so for all of his works is unfathomable. Watson concludes, "Information is the product of intelligence, not chance."

THE SCIENCE OF FINE TUNING

"There is a huge amount of data supporting the existence of God. The question is how to evaluate it."[18]

"In recent decades physicists have noticed that many of the fundamental constants of nature -- from the energy levels in the carbon atom to the rate at which the universe is expanding to the four perceptible dimensions of space-time -- are just right for life. Tweak any of them more than a tad, and life as we know it could not exist."[19]

Almost all of our science fiction and comic book fantasy assumes the existence of many other earth-like planets elsewhere in the universe. Carl Sagan and other popular twentieth century astronomers encouraged that type of thinking. In a universe of billions of stars, there must be at least some others like our sun, with planets that can support life! But, the reality may be just the opposite.

"The Rare Earth Hypothesis is the unproven supposition that although microscopic, sludgelike organisms might be relatively common in planetary systems, the evolution and long-term survival of larger, more complex, and even intelligent organisms are very rare . . . It appears that Earth got it just right! . . . This is a fate not foreseen by

---

[18] *Jewish Action*, Spring, 1999, "The Anthropic Principle," by Professor Nathan Aviezer, quoting George Ellis, University of Cape Town, p.11

[19] Ibid.

Hollywood -- that we may find nothing but bacteria, even on planets orbiting distant stars . . . We may be alone after all."[20]

Life as we know it is unique and depends on a large list of conditions seemingly impossible to duplicate. For example, while in the most recent 10% of Earth's lifetime land animals have developed, the Earth had to maintain liquid water for the previous *four billion years* to allow complex life to evolve.

Along those lines, it is interesting that Earth, Venus and Mars were all long ago covered with water. But only Earth was just the right distance from the Sun to avoid evaporation and yet be close enough to keep the oceans from freezing. Similarly, if the earth were closer to the Sun, changing surface temperatures would destroy our life-sustaining atmosphere. In addition, a magnetic field protects the Earth from cosmic radiation. Carbon dioxide making up one-third of one percent of the atmosphere is the key to maintaining a moderate surface temperature. That specific quantity is completely dependent on a natural cycle involving volcanoes and plate tectonics.

Freeman Dyson suggested that "As we look out into the universe and identify the many peculiarities of physics and astronomy that have worked together for our benefit, it almost seems as if the universe must in some sense have known that we were coming."[21]

"If the rate of the universe's expansion one second after the Big Bang had been smaller by even one part in a hundred thousand million million, the universe would have recollapsed before ever reaching its present size . . . Finally, it must be apparent that our daily survival depends on the amazing coincidence that none of the billions of meteorites, planets, and stars flying around our universe crashes into the Earth."[22]

Jupiter is 318 times more massive than Earth. Its gravity guards earth from comets and meteors. If there were no Jupiter, Earth might be hit every 10,000 years or so by a meteor like the one that wiped out the dinosaurs 65 million years ago.

---

[20] *Rare Earth: Why Complex life is Uncommon in the Universe*

[21] *Jewish Action*, Spring, 1999, "The Anthropic Principle," by Professor Nathan Aviezer quoting Freeman J. Dyson, Advanced Institute for Study at Princeton University, p.11

[22] *Permission to Believe,* pp. 50-51

The Philosophical Quest

In fact, according to L.W. Alverez, "From our human point of view, that impact (the giant meteor that crashed into the earth destroying all dinosaur life) was one of the most important single events in the history of our planet. Had it not taken place, the largest mammals alive today might still resemble the rat-like creatures that were then scurrying around trying to avoid being devoured by dinosaurs."[23] And that impact was just the right strength to destroy the dinosaur but leave other life intact.

"The probability that all of these natural forces would randomly settle at the precise strengths necessary to sustain a biotic environment is infinitesimal, as is the probability that these forces would maintain their perfect alignment every moment of the last thirteen billion years."[24]

Professor Francis Crick, an irreligious man credited with the co-discovery of DNA, said, "The origin of life appears to be almost a miracle, so many are the conditions which would have had to be satisfied to get it going."[25]

And while evolution can explain biological design over time, this explanation cannot apply to the physical laws of nature, like certain mathematical constants and balances that must exist for life to exist.

"Roger Penrose of Oxford University has calculated that the odds of the Big Bang's low entropy[26] condition existing by chance are on the order of one out of 10 to the 10(123). Penrose comments, 'I cannot

---

[23] *Physics Today*, July, 1987, L.W. Alverez, p. 33

[24] *Permission to Believe*, p. 51

[25] *Scientific American*, February, 1991, "In the Beginning . . ." J. Horgan, p. 109

[26] Entropy is a quantitative measure of the number of ways that a thermodynamic system may be arranged, or the unavailability of energy in a closed thermodynamic system. It is the measure of a system's thermal energy per unit of temperature that is unavailable for doing useful work. Work is the result of ordered molecular motion, so entropy is a measure of its disorder. Ludwig Boltzmann explained in the 19th century that an object can have a microstate and a macrostate. The macrostate of a stick of butter appears consistent in temperature and other features. But its microstate is described by the velocity and position of every atom in the bar, and there are many microstates that would describe this single bar of butter, though no onlooker would notice the difference. Entropy is the number of microstates that will correspond to that single macrostate. The higher the entropy, the more ways that there are to arrange its atoms.

even recall seeing anything else in physics whose accuracy is known to approach, even remotely, a figure like one part in 10 to the 10(123)."'[27]

Many point to this as one of two foundational mathematical relationships that support the teleological proof for fine tuning and a purposeful designer.

A formula developed by physicists Jacob Beckenstein and Stephen Hawking results in the number 1/10 to the 10 to the 123rd power representing the likelihood of our universe being created by chance.[28] Their discussion includes other numbers that are germane to the teleological idea of the Anthropic principle (creation and development for the benefit of mankind). "The fine-tuning of the universe is shown in the precise strengths of four basic forces. Gravity is the best known of these forces and is the weakest, with a relative strength of 1. Next comes the weak nuclear force that holds the neutron together. It is 10^{34} times stronger than gravity but works only at subatomic distances. Electromagnetism is 1,000 times stronger than the weak nuclear force, and the strong nuclear force, which keeps protons together in the nucleus of an atom,[29] is 100

---

[27] "Time-Asymmetry and Quantum Gravity" in Quantum Gravity 2, Oxford: Clarendon, 1981, p. 249, Craig quoting Penrose

[28] Quantum indeterminism either means that there are hidden causes or that there are none at all. But if chance is the only factor, then it's difficult to understand why the collective behavior of quantum activities displays statistical regularities. If chance is not the explanation, then all systems are continuously participating in an intricate network of causal interactions and interconnections at many different levels. Determinism must then be the result of consciousness. The conscious interaction with quantum activities is the determining cause. See the section on free will.

[29] An atom contains a nucleus consisting of protons and neutrons. It is surrounded by lighter electrons. Electrical forces keep the peace: protons are positively charged and electrons negatively charged. The 1930s saw the discovery of other particles. A positron is identical to the electron but has a positive charge. The muon is like the electron but is 207 times heavier and can be either positive or negative. Muons decay quickly and then become either electrons or positrons. The neutrino has no charge, travels at close to the speed of light and can penetrate barriers. By the 1960s so many other particles had been detected that they were given letter and number designations. Murray Gell-Mann and George Zweig proposed that protons and neutrons were composed of three smaller particles each, called quarks, which come in different "flavors" depending upon their electrical charges. They are designated as up or down. And there are anti-quarks for each quark flavor. So, there are now three levels of atomic structure: atoms comprising the nucleus and electrons, the nucleus made of protons and neutrons, and the protons and neutrons made of quarks.

times stronger yet. If even one of these forces had a slightly different strength, the life-sustaining universe we know would be impossible."[30]

Darwin thought that all life, including humans, arose from a one-celled organism. But to get from a one-celled organism to a human being with a least a trillion cells, there would have to be many changes. Darwin says these changes were produced at random, but they would have had to occur in the right order. It doesn't do any good to give an organism a leg until it has a nervous system to control it. Even if we limit the number of necessary mutations to 1,000 and argue that half of these mutations are beneficial, the odds against getting 1,000 beneficial mutations in the proper order is 2^{1000}. Expressed in decimal form, this number is about 10^{301}. 10^{301} mutations is a number far beyond the capacity of the universe to generate. "Thus, the chance of getting 1,000 beneficial mutations out of all the mutations the universe *can* generate is 10^{139} divided by 10^{301}, or 1 chance in 10^{162}." For Darwin's theory to have a chance of being right, the universe would have to be a trillion quadrillion quadrillion quadrillion quadrillion quadrillion quadrillion quadrillion quadrillion quadrillion quadrillion times older than it is."[31]

---

[30] If gravity were slightly stronger, all stars would be large, like the ones that produce iron and other heavier elements, but they would burn out too rapidly for the development of life. On the other hand, if gravity were weaker, the stars would endure, but none would produce the heavier elements necessary to form planets. The weak nuclear force controls the decay of neutrons. If it were stronger, neutrons would decay more rapidly, and there would be nothing in the universe but hydrogen. However, if this force were weaker, all the hydrogen would turn into helium and other elements. The electromagnetic force binds atoms to one another to form molecules. If it were either weaker or stronger, no chemical bonds would form, so no life could exist.

Finally, the strong nuclear force overcomes the electromagnetic force and allows the atomic nucleus to exist. Like the weak nuclear force, changing it would produce a universe with only hydrogen or with no hydrogen. In sum, without planets, hydrogen, and chemical bonds, there would be no life as we know it. Besides these 4 factors, there are at least 25 others that require pinpoint precision to produce a universe that contains life. Getting each of them exactly right suggests the presence of an Intelligent Designer.

[31] There are about 1080 elementary particles in the universe. The fastest they could mutate would be Planck time, or 10 to the minus 42 seconds. Planck time is the smallest unit of time and can be approximated as the time it would take two photons traveling at 186,000 miles per second to pass each other. If every particle in the universe (1080) had been mutating at the fastest possible rate since the Big Bang about 15 billion years ago, or 10 to the 17th seconds ago, it would produce 1080 x 10 to the 42 x 10 to the 17 or 10 to the 139 mutations. But to have a chance at even 1,000 beneficial

A second foundational mathematical relationship supporting the teleological idea is Euler's number. Represented by the symbol e, it is an irrational number, like Pi, has an infinite number of digits to the right of the decimal point (2.718281828 . . .). It is especially useful in determining growth, decay, compound interest, statistical "bell" curves, probability and a great deal more. This number is equal to -1, so when the formula is written $e^{\pi i}+1 = 0$, it connects the five most important constants in mathematics (e, π, i, 0, and 1) along with three of the most important mathematical operations (addition, multiplication, and exponentiation). "Because of the serendipitous elegance of this formula, a mathematics[32] professor at MIT, an atheist, once wrote this formula on the blackboard, saying, 'There is no God, but if there were, this formula would be proof of his existence.'"[33]

A. Cressy Morrison, former President of the Academy of Sciences of the United Kingdom, described many chance happenings in his book *Man Does Not Stand Alone*. He talks about protection from millions of meteors, ozone layer screening of the sun's rays, the oxygen-carbon dioxide exchange, the 24 hour revolution of the earth, the incline of

---

mutations takes 10 to the 301 tries. Thus, the chance of getting 1,000 beneficial mutations out of all the mutations the universe can generate is 10 to the 139 divided by 10 to the 301, or 1 chance in 10 to the 162 – a very large number! See Rich Tatum, tatumweb.com, March 10, 2006.

[32] "In the first century BCE there was a revival of Pythagoreanism. . . . Mattéi describes the Pythagorean arithmetical-theological speculations of Nicomachus of Gerasa (100 CE). Mattei argues that the Greeks did not study man and the divine in terms of 'persons,' but, rather, in terms of 'measure.' And the supreme measure of the divine is Number, in so far as it repeats itself all over the universe with its unchangeable properties. Nicomachus describes the development, the manifestation of the divine in terms of ten steps, corresponding to the numbers one through ten, that each characterizes an essential phase. One is the Monad, the supreme god, associated with Zeus, the ruler of the gods on Mount Olympus, and with Hestia, the goddess of the hearth. Two is the Dyad, the power of division and multiplication, associated with, among others, Isis, the Egyptian goddess of nature, Demeter, the giver of grain, and Rhea, that is mother earth, Gaia. Three is the Triad, representing the opposite of the Dyad, composition, and associated with a long list of gods. In this way the numbers one through ten, associated with the gods, are used to describe a ladder connecting heaven and earth that descends from the Monad, the origin of all things, to the world we see around us. (*Mathematics and the Divine: A Historical Study*, p. 17)

[33] *Christianity Today*, Charles Edward White, "God by the Numbers," 3/2006 web posting.

its axis and its distance from the sun making it habitable, the oxygen content of the atmosphere, and much more. He takes his examples from every scientific field. Morrison sums up that there isn't a single chance in millions that all the precise combinations of things necessary for life could exist without the direction of a purposeful Creator.

The odds of existence as we know it being a product of chance are incomprehensively great. The defense of chance says that in the multiverse (an infinite number of universes branching off from the original) there are many finely-tuned universes and we happen to live in one of them. But Ockham's Razor tells us to look for the simplest answer, i.e., not to multiply entities beyond necessity. Here an infinite number of universes is required to explain the fine-tuning of our own – overkill to the extreme. And, the multiverse idea is anything but simple.

There is also something called the fine-structure constant, represented by the symbol alpha in the mathematics of physics. It has been called a "magic number" whose value is "one of the greatest damn mysteries of physics" according to Nobel laureate Richard Feynman. It characterizes the strength of the force between electrically charged particles. If the value of that constant varied by just 4%, stars would not be able to synthesize carbon and oxygen atoms and carbon-based life would not exist. Yet, in a paper by John Webb and Julian King,[34] evidence is presented that the constant may not be constant after all. If true, the implications are that the universe may be even larger than we imagine and that the laws of physics vary within it. "Instead of the whole universe being fine-tuned for life, then, humanity finds itself in a corner of space where, Goldilocks-like, the values of the fundamental constants happen to be just right for it."[35] Webb found in two separate studies that alpha is varying through space. If confirmed, it may be that physical laws are not the same everywhere in the universe at all times.

Does that negate the fine tuning argument? On the contrary, it seems to simply point to the human corner of the universe as being singled out for fine tuning.

---

[34] *Physical Review Letters*, John Webb and Julian King, University of New South Wales, Australia

[35] *The Economist*, September 4, 2010, p. 85

So philosophy and religion each have a take on the divine. But the defenders of reason wage an unrelenting attack on both the idea of God and the institutionalized religions that worship Him.

II

THE REASONED ATTACK ON RELIGION

THE OPIUM OF THE PEOPLE?

When asked by the Spanish newspaper, *El Mundo*, what was meant by his statement in *A Brief History of Time* that a unifying theory of science would allow us to "know the mind of God," Stephen Hawking explained, ". . . we would know everything that God would know, if there were a God. Which there isn't. I'm an atheist."[36]

A leading light in theoretical physics, Hawking would be loath to acknowledge that religion and science are akin to twins separated at birth. They spring from the same DNA, while occupying competing realities. The brother who went to the finest private schools, camps and universities pities the lowly sibling who found his intellectual progress economically and environmentally stymied.

The scientist styles himself a humanist who believes in the agency of mankind to deal rationally with existence. He rejects the idea of a supernatural creator (or creators) who intervenes in our reality. A humanist must be by definition an atheist. Yet history demonstrates that an atheist is more accurately defined as a very "bitter" humanist – bitter about the way he has been dismissed by institutional believers (the lowly sibling) who allow their lives to be directed by irrational faith and

---

[36] CNET, 9/26/2014, "Stephen Hawking makes it clear: There is no God," Chris Matyszczyk

those who promulgate it. Bitter humanists do not simply defend their philosophical humanism, but attack theists in necessarily condescending language.

Karl Marx gave voice to the "theology" of his fellow intellectuals. "Religion is the sigh of the oppressed creature, the heart of a heartless world, and the soul of soulless conditions. It is the opium of the people."[37]

Not long ago the administration of Harvard University was forced to withdraw a proposal to revise the core curriculum to include the mandatory study of what it called reason and faith. "It's like having a requirement in astronomy and astrology. They're not comparable topics," said Professor Steven Pinker.[38] "There is an enormous constituency of people who would hold that faith and reason are two routes to knowledge. It is a mistake to affirm that."[39] And, in a 2006 study of the religious beliefs of science professors at elite universities, SUNY Buffalo sociologist Elaine Howard Ecklund found many to be infuriated by what they see as a widespread erosion of belief in proven scientific theories, such as evolution.[40] The lowly sibling must be put in his place.

Our intelligentsia has assumed the responsibility to delegitimize scripture as well as religious thought and practice. In their minds, they are performing a public service. The denigration of tradition and belief accelerates with each biological, mathematical and chemical revelation about the workings of our bodies and the universe.

Our greatest minds concur with this line of thought.

As a curious young boy Albert Einstein was quick to conclude ". . . I soon reached the conviction that much in the stories of the Bible could not be true. The consequence was a positively fanatic orgy of free thinking coupled with the impression that youth is intentionally being deceived by the state through lies; it was a crushing impression . . . The

---

[37] *Deutsch-Franzosische Jahrucher*, 1844, Karl Marx

[38] Harvard College Professor and Johnstone Family Professor, Department of Psychology, Harvard University. Author of *The Blank Slate, How the Mind Works, The Better Angels of Our Nature*, among others.

[39] *Newsweek*, "Why Harvard Students Should Study More Religion," 2/10/10, Lisa Miller

[40] *Newsweek*, "BeliefWatch: Harvard's Fuss Over Faith," 1/22/2007, Lisa Miller

word of God is for me nothing more than the expression and product of human weaknesses, the Bible a collection of honorable but still primitive legends which are nevertheless pretty childish."[41]

Whether criticized by scientists, philosophers or literary analysts, biblical fiction, perhaps fabricated by one or more authors, may in itself be sufficient to discourage belief in the God of scripture as either the creator or catalyst for events that followed. It doesn't take an Einstein to recognize that a cavalcade of modern revelations has been enough to negate any behavioral imperative that might be born of the events of that narrative.

The modern paganism of the twenty-first century's new atheists/bitter humanists is really nothing new at all. It smacks of the very religion that these skeptics so enthusiastically demean. Its foundation is the philosophy of the medieval church nominalist, a philosophy that has metamorphosed into what is known today as scientific "reductionism." The Scholastic approach of the Catholic church of the Middle Ages summed up the imperative of the faithful: all claims about the world's origin and conduct must be substantiated by meaningful and supportable explanations born of revelation, experience, insight or logic. Only the belief in divine revelation (rather than scientific revelation) as a "proof" separates the religion of the new atheist from that of the Catholic scholar.

THE UNRELENTING ATTACK ON INSTITUTIONAL RELIGION

The contemporary guardians of truth have assumed the high ground in vying with the defenders of faith, and in doing so often go after the low hanging fruit: "Despite a full century of scientific insights attesting to the antiquity of life and the greater antiquity of the earth, more than half the American population believes that the entire cosmos was created 6,000 years ago," wrote Sam Harris. "This is, incidentally, about a thousand years after the Sumerians invented glue . . . This is embarrassing. But add to this comedy of false certainties the fact that 44 percent of Americans are confident that Jesus will return to Earth

---

[41] Letter from Einstein to Eric Gutkind in 1954 in response to his receipt of Gutkind's book, *Choose Life: The Biblical Call to Revolt*.

sometime in the next 50 years, and you will glimpse the terrible liability of this sort of thinking."[42]

The attack on God and the institutionalized religions that claim Him is more than just academic, it is very personal.

"My life has been remarkably happy, perhaps in the upper 99.99 percentile of human happiness," said Nobel Prize winner Steven Weinberg, "but even so, I have seen a mother die painfully of cancer, a father's personality destroyed by Alzheimer's disease, and scores of second and third cousins murdered in the Holocaust. Signs of a benevolent designer are pretty well hidden."[43]

Explains Richard Dawkins: "There is just no evidence for the existence of God. Evolution by natural selection is a process that works up from simple beginnings, and simple beginnings are easy to explain."[44] Dawkins concludes, "The question of whether there exists a supernatural creator, a God, is one of the most important that we have to answer. I think that it is a scientific question. My answer is no."[45]

In a talk given at the Conference on Cosmic Design of the American Association for the Advancement of Science in Washington, D.C., Weinberg said, ". . . I learned that the aim of this conference is to have a constructive dialogue between science and religion. I am all in favor of a dialogue between science and religion, but not a constructive dialogue. One of the great achievements of science has been, if not to make it impossible for intelligent people to be religious, then at least to make it possible for them not to be religious. We should not retreat from this accomplishment."

Richard Dawkins, Daniel Dennett, Sam Harris, Christopher Hitchens, Steven Pinker, Noam Chomsky, Stephen Hawking, and Steven Weinberg are among those in the physical and social sciences who make

---

[42] *Newsweek Magazine*, Sam Harris, November 13, 2006, p. 42.

[43] Professor Steven Weinberg, winner of the 1979 Nobel Prize in Physics, published in PhysLink.com. Based on a talk given in April, 1999 at the Conference on Cosmic Design of the American Association for the Advancement of Science in Washington, D.C.

[44] Salon.com, in an interview with Richard Dawkins, April 30, 2005.

[45] *Time Magazine*, Richard Dawkins, author of *The God Delusion* and *The End of Faith.*, November 13, 2006, p. 51.

three basic claims. First, the Bible is a collection of fictional stories, of human origin, the product of many authors. Second, religion cannot have validity because, contrary to its stated purpose, it is the world's single greatest cause of war, death and persecution. Finally, religious claims are factually and rationally negated by science. Weinberg says confidently, "We have not yet understood consciousness or intelligence, but there is no reason to suppose that these are anything but the workings of physics and chemistry within the brain. As far as we can tell, there is nothing in the fundamental laws of nature that will suggest any special role for life or intelligence in the plan of things."[46]

And, he adds, "I think in many respects religion is a dream - a beautiful dream often. Often a nightmare. But it's a dream from which I think it's about time we awoke. Just as a child learns about the tooth fairy and is incited by that to leave a tooth under the pillow - and you're glad that the child believes in the tooth fairy. But eventually you want the child to grow up. I think it's about time that the human species grew up in this respect."[47]

When religious skeptics level their criticisms they first go after monotheistic institutional theology. Making a case for the irrational absurdity of religion requires little knowledge of antiquity, historic context or language. Christianity and its tenets offer little resistance to such criticism. Many verses in the Qu'ran are overt expressions of man's inhumanity to man. Judaism is little more than a skeletal precursor to its better developed offspring. As far as the skeptic is concerned, no distinction need be made between the violence encouraged by the God of the Torah/Pentateuch and that of the prophet in the Qu'ran.

Were the skeptic to allow himself an objective analysis, he might see that the Judaic laws and behaviors that he denigrates are historically consistent with the revealed purpose of scripture for its target audience. Without the benefit of language (Hebrew) and historical context, critics ignore all biblical imagery and allusion and seek an exclusively literal and contemporary understanding of the text. The critical arguments presented against faith and their biblical texts are so simplistic that were

---

[46] 1999 PBS radio interview with Steven Weinberg, posted on the PBS web site
[47] Ibid.

comparable critiques offered in academic fields, their proponents would be savaged by expert colleagues.

The astronomical beliefs attributed to the Old Testament were given to a humankind that critics unreasonably demand should have been scientifically sophisticated 3300 years ago. Animal sacrifice, slavery and incest are seen in only their literal sense, not in the context of weaning a primitive, pagan mankind from inhumane practices and beliefs. Behavioral evolution is not inferred from narrative.

And they would vigorously dispute the believer's explanation of creation that "In a free-willed act an all-powerful being designed and created a natural universe containing entities that are morally responsible for their choices."[48] Instead, the man of science insists that, "The aim of physics, or at least one branch of physics, is after all, to find the principles that explain the principles that explain the principles that explain everything we see in nature, to find the ultimate rational basis of the universe."[49]

If rationality is the measuring stick, then everything must be contained within the system. There can be no reason to look beyond. "Elementary particle physics is different," says Dr. Weinberg. "Its aim is to complete our understanding of the ultimate rules that, at bottom, govern everything."[50] The reality that we observe must be the product of chance occurrences operating within a set of natural rules, which themselves exist due to a chance combination of events at creation. There need not have been someone or something in charge. What we see around us is purposeless. It just is.

Nobel Prize winner Weinberg explains existence in just that way: "If we were to discover that the laws of nature at their deepest level give some special role to life or intelligence, we might derive from this some sense of purpose, some standard of conduct. My own guess is that this will not happen . . . My own feeling is that there will be something

---

[48] B'Or Ha'Torah, No.13, p.7, Avi Rabinowitz, PhD, "And God Said, Let There Have Been a Big Bang"

[49] 1999 PBS radio interview with Steven Weinberg, posted on the PBS.org

[50] Ibid.

healthy in coming to grips with the knowledge that we, like the rest of nature, are the way we are because of impersonal natural laws."[51]

But, the insufficiently informed skeptic actually goes a step further and sees the influence of religion as destructive to humanity! Richard Dawkins refers to this "faith" as a disease that acts like a virus, making the brain do the irrational. Corey Fincher and Randy Thornhill, biologists at the University of New Mexico, took that thought a step further.[52] They concluded that people actually develop "behavioral" immune systems that allow them to be exposed to and develop resistance to bacteria and other organisms that would adversely affect those from a different environment. They believe that religion, by closing off the individual or group from broad exposure, is a response to the threat of disease. So religion does, in fact, protect the believer from harm, but not for the reasons that its practitioners believe. Outsiders may carry pathogens lethal to this formerly underexposed group. Their study found that patterns of behavior that promote group exclusivity (primarily organized religious behavior) are stronger in disease ridden areas.

Disbelievers routinely deliver empirical arguments against traditional religious claims:

For example, are "miracles" really the acts of a supernatural being? We often think of a miracle as a one-in-a-million event. If so, then the miraculous is statistically more common than imagined. According to "professional skeptic" Michael Shermer, we can reasonably estimate that in the course of a typical day, the flow of data into our senses is about one "bit" of information per second. In 12 waking hours then, we will receive 43,200 bits of data or 1.296 million bits of data in a typical month. If 99.9% of these bits of data are meaningless, we are still left with what we might think of as 1.3 miracles a month, or 15.5 a year! [53]

---

[51] Ibid.

[52] *Behavioral and Brain Sciences*, 2012, April ; 35(2): 61-79., C.L. Fincher, R. Thornhill. According to published data, the average number of religions per country is 31 and varies from 6 to 643. For example, Brazil has 159 and Canada 15. There on average 200 parasitic diseases per country, a range of 178 to 248. The number of religions in a particular place seems to correlate closely with the share of the population carrying disease.

[53] *Scientific American*, Michael Shermer, Skeptic,, November 2010, p. 86

Darwin used the "miracle" of the exquisitely complex human eye to make an argument for the "fine tuning" of creation (sometimes called the Teleological Proof for the existence of God): "To suppose that the eye, with all its inimitable contrivances for adjusting the focus to different distances, for admitting different amounts of light, and for the correction of spherical and chromatic aberration, could have been formed by natural selection, seems, I freely confess, absurd in the highest possible degree."[54]

Michael Shermer again offers convincing skepticism: "If God created the eye, then how do creationists explain the blind salamander? . . . Why on earth would God create a salamander with vestiges of eyes? If he wanted to create blind salamanders, why not just create blind salamanders? Why give them dummy eyes that don't work and that look as though they were inherited from sighted ancestors?"[55]

The idea of divine design or fine tuning is routinely debunked by the philosophers of science: "We speculate a lot about things we see as fundamental, like the masses of the particles, the different varieties of forces, the fact that we live in three space dimensions and one time dimension. But maybe all this is not fundamental but environmental. The universe may be much more extensive than we've imagined, with much more than just the big bang that we see around us. There may be different parts of the universe – where parts could mean various things – that have very different properties and in which what we normally call the laws of nature may be different and even the dimensionalities of space and time are different. There has to be some underlying law that describes the whole thing, but we may be much further from it than we now imagine. "[56]

Doubters apply both "unassailable" reason and natural science to make their case against faith: "Two-fifths of Americans still refuse to accept that human beings share a common ancestry with animals, preferring to believe that they were created in their present form in the past 10,000 years."[57] Or, "Surely an intelligent designer would have put

---

[54] *The Origin of Species*, Charles Darwin, J.M. Dent & Sons Ltd, London, 1971, p. 167

[55] *Scientific American*, Michael Shermer, Skeptic, November, 2010, p. 86

[56] *Scientific American*, Nov. 2010, p. 67

[57] *The Economist*, Sept 5, 2009, p. 90.

the rainforest canopy somewhat lower, and saved on tree trunks? The cheetah is perfectly honed to hunt gazelles – but the gazelle is equally well equipped to escape cheetahs. So whose side is the designer on? The ichneumon wasp paralyses its prey without killing it and lays its larva inside this convenient source of fresh meat, to eat it slowly alive. This is just one striking instance of the immensity of pain in the animal kingdom, which defies explanation except via the unyielding calculus of competitive survival."[58]

Even the association of religion and morality is questioned. Sam Harris rebuts the claim that religion is required for a moral society by stating that morals and values are based on the way things are: neuroethics says the well-being of conscious creatures comes from our building a science-based system of moral values by quantifying whether or not something increases or decreases well-being.[59]

Christopher Hitchens elegantly shuts down any argument for divine design. "It's only once we shake our own innate belief in linear progression and consider the many recessions we have undergone and will undergo that we can grasp the gross stupidity of those who repose their faith in divine providence and godly design."[60] "Creationists begin with answers and work to prove that those answers are right. This is antithetical to the scientific process."[61]

---

[58] *Ibid.*

[59] *The Moral Landscape*

[60] Slate.com, Christopher Hitchens, "Losing Sight of Progress," 7/21/2008

[61] *Scientific American*, Jacob Tanenbaum, Forum, January, 2013, p. 11

III

THE RESPONSE OF SCRIPTURE: THE CHOSEN CONCEPT

THE MANIPULATIVE POWER OF ALTERNATIVE FACTS

Author, columnist and professional skeptic, Michael Shermer was asked,[62] "If you were presented with credible archaeological and scientific evidence in support of the biblical narrative, would that have any influence on your attitude toward religious faith in the God of the Bible?" He replied: "There is already a lot of archaeological evidence in support of many parts of the biblical narrative, and virtually none for other parts. But none of that evidence speaks to the veracity of supernatural claims related to biblical stories. For example, there may well have been someone named Abraham, Moses or Jesus who lived, but evidence for their lives says nothing about whether or not they spoke to God, received the commandments from God through a burning bush, or was resurrected after being dead for three days. Further, I know of no way to prove such supernatural claims using the natural sciences."

Connecting the biblical narrative and the divine is further discouraged by evolutionary biologist and author Richard Dawkins. "I suppose the most strident passage in *The God Delusion* is where I talk about how the God of the Old Testament is the most unpleasant character in all fiction. I had this long list of adjectives: homophobic, infanticidal. That's kind of using long words, long Latinate words to describe what

---

[62] Personal email exchange in March, 2009.

everybody actually knows: that the God of the Old Testament is a monster. I put it in this rather, I'd like to think, amusing way."[63] When told that 90% of Americans believe in God, Dawkins insisted that the intelligent among them wouldn't disagree with his assessment of the God of the Old Testament. They would "probably say something like, Oh, that's quite different. We believe in the God of the New Testament." When it was suggested that Jews would not answer in that way, Dawkins said, "Well, sure enough. They'd say, 'OK, we've moved on since that time.' Thank goodness they have."

As in many other walks of life, confirmation bias is hard at work here. If skeptics see religion as scientifically indefensible, it is due in large part to their pointed resentment of fundamental Christian belief and practice. They rail at the irrationality of scripture and at those who profess fealty to it. Yet neither they nor their Christian provocateurs have fluency in the language of the original scripture. We will see a little later that what is known today to the Christian world as the Old Testament is an altered facsimile of the original, burdened with an agenda that exists to retroactively support a fantastical, follow-on history that is nowhere contemporaneously corroborated.

So, the skeptics are correct in their attack on what they see as monotheism, but for the wrong reasons. They have taken at face value a Greek and then Christian revision of the Hebrew Bible retrofitted with a "New" Testament and correctly identified it as an indefensible body of belief. The original monotheistic faith is barely addressed, and when it is, only in the most superficial and astonishingly infantile manner.

A two-pronged offensive was launched against biblical religious belief and practice two centuries ago, aimed first at its historical accuracy and second at its claims of divinity. Critics come from both the worlds of science and the arts. Scientists base their attack on what appears to be the obvious inaccuracy of the biblical account of creation – its impossible timeline, claim of spontaneously generated, fully-formed life and an apparent frontal assault on cosmologically accepted, scientifically supported creation theories. They have concluded that it can be nothing more than a faith-based attempt to ascribe divinity to the starting point for existence as we know it.

---

[63] *Newsweek*, 10/5/2009, interview of Richard Dawkins by Lisa Miller, p.54.

Experts in language and the philosophical arts base their objections on the apparent inaccuracy of biblical history and the literary inconsistency of the biblical narrative. They also continue to call attention to a lack of necessity to ascribe any historical events or consequences to powers other than humanism and chance.

If there is to be a convincing rebuttal to the skeptic, then it must be based on three principles: first, the unlikely persistence of the Jews; then, an explication of the tools provided by the Hebrew Bible, demonstrating that the creation account is scientifically accurate; and finally, the demonstrable historical validity of the subsequent biblical narrative. An accurate historical narrative that depicts the consequences of actions that must be initiated by a source beyond the bounds of nature can go a long way toward reducing or eliminating scientific doubt.

We begin with the introduction of a fundamental biblical principle – the analysis of which invites a deep examination and understanding of the process of creation and the link between text and science.

Why all the fuss about a "Chosen" people?

Why get excited about the biblical idea of a "chosen people?" The world has hundreds of distinct ethnicities and more than 6000 languages. If one numerically insignificant group of people wants to think of itself as special, why should the rest of the world object? But we must be missing something here because there continues to be an unrelenting struggle to claim and retain that title. It seems that the approbation of a God that most enlightened humans no longer believe exists is worth fighting and dying for.

Contenders for the title see a tiny "nation" to whom the "chosen" title originally belonged defying both nature and history. The Jewish people persist like a sore on the rump of a pig – forever bothersome enough to scratch, slap and rub into the mud, but impossible to remove. Neither local nor industrial-scale plans for its extermination have succeeded. The Jews have survived a succession of Holocausts -- each an attempt to cleanse a state, a country or the world of God's chosen people, the most venal race ever created. And the two most populous monotheistic religions of the world fight bitterly to inherit that "chosen" crown.

War, disease, assimilation and natural disaster should certainly have reduced the Jewish people to little more than an historical footnote. Instead, history testifies to disproportionate Jewish influence in every significant walk of life – science, literature, politics, medicine, commerce, philosophy. Is there something more to this "chosen" designation than social or ethnic pride? And if there is, what does that say about the role of this special people in course of history?

Can the biblically publicized claim of Jewish exceptionalism be validated?

Invoking the philosopher's logic might generate this "proof:"

- The Jews have a long and exceptional history of survival and accomplishment.
- There must be a reason for that exceptional history.
- A reason is given in the Torah/Five Books of Moses. (Of course, many will argue that the reasons are sociological or circumstantial – but we'll address that later.)
- If the Torah is true, the reason has a supernatural explanation.
- If the Torah is true – if science and history support that fact -- then God exists.
- (modus ponens): The Jews are exceptional for divine reasons.

Let's follow this reasoning a little further: If the skeptic accepts the initial assertion (that the Jews have an exceptional history) and the logic that follows, then he must prove a negative. Logic dictates that those who would deny that Jews are objectively exceptional for divine reasons must prove at least one of the following three assertions:

- Jews do not differ experientially, physically or spiritually from anyone else (they are not "naturally" exceptional).
- God does not exist so there can be no supernatural explanation for Jewish exceptionalism (they are not "supernaturally" exceptional).

- The Torah is not divine, and therefore cannot be the source of a divine designation of Jewish exceptionalism (their idea of exceptionalism has no "third party" historical validation).

For more than two centuries, the effort to disprove this logic has concentrated on disproving the divinity of the Torah/Pentateuch. In order to do so, the content of both history and scripture have been overturned and reformatted to meet the needs of science, literature and competing religious narratives.

BEING CHOSEN

There is no more controversial or misunderstood religious or sociological concept than that of the "chosen people." Its mere mention evokes resentment, arrogance, anger or embarrassment. Who has been chosen, by whom and for what? And what is the status of that chosen group vis-à-vis its neighbors?

The chosen designation has been associated with the Jewish people from time immemorial. It was conceived for them (or many will say, by them) and only them. They lay claim to a scriptural, divine designation and an assignment of unusual responsibility. Successful implementation of a divine plan built around that designation is promised to have global implications. Consequently, for thousands of years a succession of other religions and nations has sought this mantle, and in each case, that claim has coincided with an historic political or religious paradigm shift, changing the very face of contemporary society.

The skeptic correctly sees religion as a primary cause of war and genocide -- a man-made construct separating natural brothers. He barely notices Judaism. He sees nothing there that distinguishes it from far more influential religions. In the process of equating all religions, there is no reason to remark the now diminished biblical status originally conferred on the Jews. And, of course, if there is no God, then scripture is fiction and chosen status is meaningless.

But what if the skeptic is wrong? What would history look like with the Jewish people embracing the exceptional role granted by scripture? Would their internecine quarrels have been any less common or their external persecutions any less frequent? Would the nations of the world have related differently to Jews and to each other? We may be able to get a sense of where to look for answers to these questions by examining and explicating the components of divine exceptionalism as they were originally applied to scripture's original designee. The conduct of this investigation leaves arguments against God's existence, the separation between faith and science and the claims of competing religions significantly weakened.

The salient question and its implications are straightforward: If God exists and the Torah is divine, then the Jews were chosen and promised that designation eternally. And if that designation is eternal, then it applies only to the Jews, and their responsibilities and benefits accrue accordingly. There is no declaration in scripture that minimizing or ignoring this designation will expunge or limit its obligations. On the contrary, failure to assume these designated responsibilities is a prescription for unpleasantness for the entire Jewish nation – a circumstance that will persist until accountability is restored. It is entirely logical then, that given the divine assumptions stated, the Jewish people would be subject to the depredations of their past and current "exiles" for casting off their chosen role.

So, in order to define Jewish exceptionalism and to better understand what might be expected if Jews recommit to that "chosen" role, let's begin with the basics.

THE MODERN DEVELOPMENT OF THE EXCEPTIONALIST IDEA

In the late 18th century, the German theologian, philosopher and poet, Johann Gottfried von Herder formulated and began to articulate the idea of national exceptionalism. A student of Kant and colleague of Goethe, Herder brought about a renewal of German pride in its ancient origin, language and culture. He insisted that nationality and patriotism would lay the groundwork for a classless society brought about by providence. Christianity and idealism would be manifest in the exceptional nation that he hoped Germany would become.

But, he also understood the "racist" potential that this exceptionalism implied and argued forcefully for the rights of and respect for the Jews. He was only too aware of the scriptural exceptionalism bestowed upon the Jews by God and the supersessionist claims of Christianity. And he saw quite clearly the fear of the Jew, even then, in his countrymen.

Von Herder's exceptionalist idea nurtured the political identity that visited unimaginable horrors on both the Jews and others in the twentieth century. As a result, by early in the 21st century the idea of national exceptionalism had fallen so precipitously out of favor, that even the president of the United States denied its applicability to his own country's role in the world order. That fall from grace paralleled the earlier de-emphasis of more than 1500 years of religious exceptionalism in the Christian West. The idea that one nationality, faith or culture was inherently superior to others was simply "undemocratic," counterproductive, and certainly far from the will of a just and caring God.

In keeping with this notion of universal equality, all contemporary references to Jewish exceptionalism have been proffered with arrogance and received with cynicism. The idea of a certain people being "chosen" for special consideration by the Creator seems irreconcilable with the belief that all men are the children of one God. Yet, the existence of even the smallest possibility that Jewish exceptionalist claims may have some contemporary validity is motivation enough for behavioral, political and sociological counter-measures. In a world in which universal equality is promoted (though not necessarily observed), no single people, nationality or ethnicity has a right to make such a claim for itself, let alone act upon it.

Examples of national exceptionalism are legion. Ancient Egypt was driven by its accomplishments in science, language and construction; Greece, by its academics and philosophy; Rome by its conquests, both military and spiritual; Nazi Germany by its ethnic purity. The exceptionalism of the British Empire was built on economics; that of the United States on multi-culturalism and capitalism. But providence has turned its back on each. Each has faced the self-doubt born of competitive ideologies preying on wavering spiritual and political loyalties. Ancient exceptionalism has always claimed the approval of God. From the divine right of kings to the Holy Roman Empire to the simple act of a religious invocation at a coronation or installation ceremony, religion has blessed national and ethnic aspirations with a divine, exceptional status. But, Christianity and Islam, the primary conferees of that exceptional political status, have each suffered from internal misgivings that led to the Reformation and Islamism, and ultimately a diminution in their claimed exceptional standing.

Judaism holds the original claim to religious and sociological exceptionalism, though its adherents have not fully exercised that claim for two millennia. Does Jewish exceptionalism have a place in the contemporary narrative? How do its claims compare to those of its monotheistic cousins? What role has that claim played in its past and will it play in an uncertain Jewish future?

Religious Exceptionalism

The children of Abraham are engaged in a sometimes overt competition, demanding recognition as the one faith designated by God for primacy. That contest is fought in the classroom, in the media, on the battlefield and in the streets.

The earliest recorded exceptionalist claim states that God singled out a particular people for a unique role. In doing so, He appeared to that people in a mass revelation in order to charge them with a portfolio of associated responsibilities. Thirteen centuries passed before this exceptional role was usurped by a successor claiming to make up for the repeated failures of the original designees. The mantle would be passed to a sect led by a simultaneously divine and human emissary. Six hundred years later, the vision received by a human "prophet" left all previous exceptionalist claims null and void. Only his could be the correct reading of the desires of God for man, and only the followers of his retrofitted theology could be trusted with that role.

Religious exceptionalism is a product of the biblical narrative and the chosen concept therein. According to that narrative, only one people was chosen by the Creator to observe an historically atypical moral and social standard in perpetuity. The offspring of the forefather Jacob (who was also known in scripture as Israel) were designated by God as his "am sigulah"-- typically translated as the chosen or treasured nation. This description appears four times in the Torah/Pentateuch (Exodus 19:5, Deuteronomy 7:6, 14:2, 26:18). Its subsequent use in Chronicles (29:3) and in the Babylonian Talmud (Bava Batra, 52a) makes it clear that the term implies distinct, private ownership.[64] The designation is further emphasized in the book of Leviticus when God demands that his people be "kadosh" as is He. That "separateness" is highlighted there by the language used to describe this unique people. They are not referred to as a nation, but as an "eidah," a collective witness or testimonial to this divine designation. The God of the Pentateuch and the prophets designated a specific people as His own, gave them a specific narrative of creation and worldly events and provided them a living, legal code, the observance of which separated them from all other nations of the time.

This sense of the exceptional – a sense that forced a dramatic recalibration of ancient, inter-national and inter-ethnic relationships -- was later de-emphasized in rabbinic Judaism. This, in part, encouraged the development of replacement theologies, first Christianity and later, Islam. According to these new exceptionalists, Jews had not only failed to live up to their divine role, but they had abandoned their self-identification with it. Christianity maintained that God offered a new testament presented by a physically accessible version of Himself. New teachings would replace the original theology, and discredit its ancient "select" and their heretofore special relationship with God. Islam later denigrated and subverted the very narrative on which exceptionalism was conceived in order to stake its own revisionist claim.

CONTEMPORARY EVIDENCE OF JEWISH EXCEPTIONALISM?

Mark Twain said, "If the statistics are right, the Jews constitute but one quarter of one percent of the human race. It suggests a nebulous puff of star dust lost in the blaze of the Milky Way... (yet, the Jews) are peculiarly and conspicuously the world's intellectual aristocracy... (their) contributions to the world's list of great names in literature, science, art, music, finance, medicine, and abstruse learning are ... way out of proportion to the weakness of his numbers. He has made a marvelous fight in this world ... and has done it with his hands tied behind him."[65]

---

[64] Academia.edu, "Notes on Some Hebrew Words in Ecclesiastes," Stuart Weeks, 2013

[65] Mark Twain, September, 1897, quoted in the *National Jewish Post & Observer*, June 6, 1984

Jews comprise 0.2% of the world's population, yet in the 20th century nearly 20% of all Nobel laureates were Jewish, as were 54% of the world chess champions, 27% of the Nobel Prize winners in physics and 31% of those in medicine. In the U.S. where Jews comprise about 2% of the population, 21% of Ivy League students are Jews as are 26% of Kennedy Center honorees, 37% of Academy Award winning directors and 51% of the Pulitzer Prize winning authors for nonfiction.[66] Twenty three of every 1000 Ashkenazi Jews has an IQ of 140 or greater as compared with 4 of every 1000 Northern Europeans.[67]

A 2006 study by Richard Lynn and Tatu Vanhanen (*IQ and Global Inequity*) calculated an average Jewish IQ of 115, 8 points higher than the closest ethnicity (Northeast Asians) and 40% higher than the global average IQ of 79.1.[68] In 1954, test results were used to identify 28 children in the New York public school system with IQ's of 170 or greater. Of those 28, 24 were Jewish.[69]

Ynet news reported that (May 12, 2011) South Koreans are so taken with Jewish achievement that they require their children to study Talmud! In fact, they read a collection of stories/fables translated from the Talmud, as if to imply that the learning skills that Jews have cultivated over the centuries can be reproduced in Korean youth through this association.

Cyril Darlington offered that the mold for the genetic continuity of Jewish intelligence and intellectual accomplishment was formed in Babylonia, after the destruction of the First Temple. Nebuchadnezzar "carried into exile all Jerusalem: all the officers and fighting men, and all the craftsmen and artisans . . . Only the poorest people of the land were left."[70] And those who returned after 70 years to begin rebuilding the Temple were primarily descendants of that elite group taken into exile.

---

[66] Aish.com "Endangered Jewish Genius," Rabbi Benjamin Blech, Jan 8, 2013

[67] "Natural History of Ashkenazi Intelligence," Gregory Cochran, Jason Hardy, Henry Harpending, University of Utah, web.mit.edu, 2006

[68] Aish.com, "Endangered Jewish Genius," Benjamin Blech, 2013

[69] *Commentary Magazine*, "Jewish Genius," Charles Murray, April, 2007, p. 29

[70] 2 Kings 24:10

Even though many claim that the Jews' apparent intellectual inactivity during the Medieval period disproves this contention, a detailed retelling of the history of science found that between the years 1150 and 1300, 95 of the 626 known scientists working anywhere in the world were Jews – 15 percent of the total![71] This refutes the results of a 2006 study that concluded that elevated IQ's were limited to Ashkenazim (Jews of Eastern European origin), who while living in Europe were limited in occupation to sales, finance and trade. Sarton's history demonstrates that Sephardic Jews of the Iberian Peninsula, in Baghdad or other Islamic centers filled this list, as they did later in Spain where they were prominent in medicine, government, commerce and in literary circles. Interestingly, more than 80% of the Jews at the beginning of the Common Era were farmers. But less than 100 years later, fewer than 20% were. Some trace this change to the edict of Rabbi Joshua ben Gamla in 64 C.E. mandating compulsory schooling for all males beginning at age six.[72]

Is this Jewish intellectual exceptionalism just a product of culture and history, or does it date back to a covenant between God and one people? Political scientist Charles Murray asks, "Why should one particular tribe at the time of Moses, living in the same environment as other nomadic and agricultural peoples of the Middle East, have already evolved elevated intelligence when the others did not? At this point, I take sanctuary in my remaining hypothesis, uniquely parsimonious and happily irrefutable. The Jews are God's chosen people."[73]

---

[71] *Introduction to the History of Science*, George Sarton

[72] *Journal of Economic History*, "Jewish Occupational Selection: Education, Restrictions, or Minorities?" 2005, Maristella Botticini, Zvi Eckstein

[73] *Commentary*, "Jewish Genius, "p. 35

IV

WHO ARE THE REAL CHOSEN PEOPLE?

A BRIEF HISTORY OF RELIGION

Historians and sociologists trace the origins of religion back nearly 300,000 years. This dating is based on archaeological discoveries suggesting that religious rituals were long associated with human burial practices. And while there is some controversy about just how to interpret these finds, there is almost none at all associated with "religious" artifacts dating back "only" 20 or 30,000 years.

For two centuries these religious precursors have been described by experts as everything from naturalism to animism, fetishism, and hedonism. All were indistinguishable from mythology. To ancient man, there were many explanations for existence and nature, no one of which eclipsed all others.

As people began to gather in communities during the Neolithic period (9000 years before the Common Era) formal religion began to take shape. The development of written language led directly to religious text, examples of which date back more than 4000 years to Ebla (an iconic Syrian archaeological site) and early Egypt.

The period that began in the millennium before the Common Era was called the Axial age by German philosopher Karl Jaspers. It was generally agreed that during these several centuries new ways of thinking inexplicably appeared or took root around the world. Monotheism and the major religious traditions of the Far East gained traction with the lion's share of humanity. The accepted chronology of religious introduction and development looks something like this:

2,000 BCE	Judaism
1,500 BCE	Hinduism
1000 BCE	Zoroastrianism
560 BCE	Buddhism
550 BCE	Taoism
500 BCE	Jainism
30 ACE	Christianity
50-100 ACE	Gnosticism
325 ACE	Institutional Christianity
632 ACE	Islam

Judaism, Christianity and Islam accept the idea of divine scripture, though each has its own version. The texts introduced by Christianity and Islam by their own definition supersede that of Judaism.

THE NATURE OF CHRISTIAN EXCEPTIONALISM

Christian supersessionist (replacement) theology is unequivocal: the Jews failed repeatedly to fulfill the mission with which they had been charged by God, and as a result, were replaced as His chosen people.

St. Augustine of Hippo argued in his *City of God* that Jews merit the loss of the land (of Israel) for neglect of their duties. He cites Jeremiah 31:31: "Behold, the days come, says the Lord, that I will make for the house of Israel, and for the house of Judah, a new testament: not according to the testament that I settled for their fathers in the days when I laid hold of their hand to lead them out of the land of Egypt; because they continued not in my testament, and I regarded them not, says the Lord." Augustine believed that that this promise was expressed in the New Testament and was fulfilled with the appearance of Jesus.

Romanian religious historian and University of Chicago professor, Mircea Eliade wrote, ". . . Christianity entered into History in order to abolish it: the greatest hope of the Christian is the second coming of

Christ, which is to put an end to all History."[74] The Letter of Barnabas[75] expresses the Christian understanding that "Jews misunderstood the Old Testament, which, the writer asserts, was never intended to be observed literally, since all therein is but a prefiguring of Christ and the Church."[76]

The idea that the Church replaced the Jews as the chosen of God is fundamental to Christian belief. The Second Vatican Council, generally thought of as an effort to soften anti-Jewish rhetoric, reiterated in 1965 that "the Church is the new people of God." All of the main streams of Christianity cite the New Testament verse from Hebrews 8:13 (referring there to Jeremiah 31: 31-32), "In speaking of a new covenant he has made the first one obsolete." In fact, Cardinal Joseph Ratzinger, later Pope Benedict, wrote "To imitate him (Jesus), to follow him in discipleship, is therefore to keep Torah, which has been fulfilled in him once and for all. Thus the Sinai covenant is indeed superseded."

"On the one hand the (Vatican) Council encourages dialogues with other religions; on the other, it also affirms that the Roman Catholic Church is the only repository of all true religion," said Rabbi Eliezer Berkovits. "What then is the purpose of the dialogue for the Church? . . . It can have only one purpose – to spread the good tiding to the unfortunate who have not yet seen one's own light."[77]

After the Second Vatican Council, the Nostra Aetate (1965) used the language of Paul of Tarsus: "although the church is the new people of God, the Jews should not be presented as rejected or accursed by God, as if this followed from the Holy Scriptures." Of course, if Israel has not been fully rejected, but has been condemned and replaced with the church by God, then the church must address the fact of the continued existence of the Jewish people and their reestablishment of a national

---

[74] *Faith After the Holocaust*, p. 56

[75] The Letter, or Epistle of Barnabas is an anonymous work likely written in the late second century and attributed to Barnabas, a senior companion to Paul. It is aimed at Christians who may be in danger of lapsing back into Judaism and "Judaizing" error. It shows that Christianity is the one, divine method of salvation and disparages Judaism with its various laws and practices. The new law of Jesus is given without constraint. It is sinful to believe that the covenant of the Jews is still binding. He believes that the Jews have misinterpreted the Old Testament and debased it.

[76] *Anguish*, p. 34

[77] *Faith After the Holocaust*, p. 45-46

homeland in Israel. Ultimately, though, the Christian believes that the best that the Jews can hope for is that the teaching of Revelations 20:1-7 will come to fruition. Jesus will return as king; his kingdom will last 1000 years and reign over the whole earth. Israel will then recognize him and only then be restored as God's chosen (Romans 11).

Early church fathers actively promoted this belief. Justin Martyr wrote, "For the true spiritual Israel are we who have been led to God through this crucified Christ."[78] The coming of Jesus was an acknowledgement of Jewish failure and the negation of the old rules. In Romans 7:6-8, Paul states that the law (Torah) causes sin instead of leading to holiness: "Now we are rid of the Law, freed by (Jesus') death from our imprisonment, free to serve in the new spiritual way and not the old way of a written law . . . What I mean is that I would not have known what sin was had it not been for the law. If the (Torah) had not said 'You shall not covet,' I would not have known what it means to covet . . . when there is no law, sin is dead."

Thomas Jefferson wrote his own interpretation of the Christian Bible calling it the "Philosophy of Jesus." He removed all of the mystical material but showed admiration for Jesus as a moral teacher. In his eyes, the mission of the church was to reverse the decadence of the Jewish religion.[79]

Origen[80] was the first great systematizer of early Christian philosophy. He wrote and preached that the Jews were married to the "word" (the laws of the Torah) and the physical and so they murdered their god. "Christianity was not a new religion but rather the true and ancient Israel whose prophecies were more ancient than Homer's and whose teachings understood allegorically, were not only consonant with but superior to those that philosophy could offer. Moses's laws were better than Plato's, his cosmogony more reliable. In fact, says Origen, Plato borrowed much that was good in Greek philosophy directly from the Israelites – which is to say, in Origen's schema, from Christianity, the

---

[78] Justin Martyr, Dialogue with Tryphgo 11. In Ante-Nicene Fathers 2:389. He lived about 150 C.E.

[79] *The Economist*, Dec 17, 2011, p. 35

[80] Origen lived at the end of the second century and the beginning of the third and was a philosophical interpreter of the canon.

true Israel – whose teaching he encountered during a tour that (according to an ancient tradition) he had made of Egypt."[81]

"In the mid-first century, Paul laid down the ground-plan for the Church's theology of Israel: The Law, transitory and preparatory in character, terminated in Christ. Universal salvation is found in faith in Christ, which is the fruit of grace. The burden of the Law is replaced by the hope and liberty of the Gospel." ". . . even if the Jews have failed by their unbelief, God has not cast off His people. . . In the fullness of time, they will return, and their reconciliation will be a golden age for the Church. The task of Christians is not to patronize them but to provoke them to jealousy by the holiness of their own lives."[82]

Professor David Nirenberg writes, "All (the church fathers of the first 4 centuries of the Common Era) more or less agreed that the prophets themselves, properly understood (that is, spiritually, allegorically, typologically), had never really been Jewish. Jews were those whether Jewish or Christian – who misunderstood prophecy (that is, read it literally) and were condemned for it. By insisting on a thoroughly 'spiritual' reading of the Hebrew Bible, these theologians deprived the Jews of their scriptures, and the scriptures of their Jews."[83]

Martin Luther, like the vast majority of Christians, continued to associate the Jews with the aberration of living by the law. Luther wanted to disabuse men of the belief that they deserve the remission of their sins as a reward for their own works (as rabbinic Judaism believed). He wrote that all of Hebrew scripture was about Jesus. He insisted that Jews had perverted scripture. The Psalms were not about David's suffering and faith, but about the suffering of Jesus in Judea under Herod and Pontius Pilate.

"If the literal meaning of the Hebrew Scriptures is the life of Jesus," Nirenberg continues, "then Jews lose their traditional Augustinian role as guardians and guarantors of the letter, and are cast exclusively as persecutors of Christ and of Christians."[84]

---

[81] *Anti-Judaism*, p. 105

[82] *Anguish*, p. 30

[83] *Anti-Judaism*, p. 120

[84] Ibid, p. 254

THE LIMITATIONS OF CHRISTIAN EXCEPTIONALISM

Despite doubt concerning the historicity of Christian theology, few question the Christian appropriation of the "chosen" mantle from the Jews. The Christian claim to exceptionalism is based on a belief that God separated or created a piece of Himself and visited earth in human form. That God/man absorbed the sins of all men, thereby absolving believers of the need to observe any previously given divine law. The compatriots and followers of that "son of God" wrote down the events of these heady days and described the divine nature of Jesus and his teachings.

To make a case for exceptionalism, to wrest that claim away from the Jews, Christian historians had to present a convincing narrative. There are no eye witness accounts to either the events recounted or the protagonist himself, and there is no archaeological or corroborating historical evidence to support any claims. This narrative seems to have been composed after the fact in order to bolster the creation of a new faith and the developing tenets of the soon to be institutionalized religion.

While there has been very little practical impact, later historians and clergymen have frequently questioned the religious history that is fundamental to Christian claims. "No mention is made of either of the Gospels Luke, Mark, John or Matthew, by Clement, Ignatius or Polycarp."[85] "No reference is made to either of the four gospels, nor to the Acts of the Apostles, nor are there any quotations except such as evangelical writers concede may have been taken from other sources."[86]

H.L. Menken concluded, "There is no reason, indeed, to believe that He (Jesus) was ever regarded as a god before His death, save maybe by a few extremely fanciful followers, or that He had ever heard anything about the virgin birth. The earliest Christian documents, the epistles of Paul, do not mention it, nor is it mentioned in the gospels of Mark and John or in Revelation. The references to it in Matthew and Luke are probably interpolations, for, as everyone knows, they conflict with the genealogies in the same gospels, showing that Jesus descended from

---

[85] *History of the Christian Religion to the Year 200*, p. 57

[86] Ibid, p. 61

Abraham through Joseph. All of the gospels, as we now have them, were written years after Jesus's death, and by men who had never seen Him. They show an assimilation of myths that were old before Jesus was born – some essentially Jewish, but others belonging to the common stock of religious ideas in the Near East."[87]

The first century Jewish exegete, Philo of Alexandria seems to be the source of a number of texts produced by New Testament authors. "The Logos, the Parable of the Prodigal Son, the Gift of Tongues, as well as Barabbas have all been derived from Philo. We saw that Philo spoke of the crucifixion of Karabas . . . moreover, some of the old texts of Matthew speak of Jesus Barabbas, as contrasted with Jesus, the Son of the Father in Heaven. The Aramaic for Son of the Father is Barabba. Does this not suggest that originally Jesus and Barabbas were one and the same person?"[88]

In 1909 John Remsburg (*The Christ*) composed a list of 42 ancient authors who lived contemporaneously with or shortly after Jesus and were not aware of either his existence or that of Christianity. And no later Christian writer has claimed to have read about Jesus in any text from that period.

The "proofs" that Christian historians (Strobel) have subsequently offered are citations by later authors and are sometimes doctored texts (as has been demonstrated with the reference to Jesus in the *Antiquities* of Josephus). There are no recordings of miracles, crucifixion, resurrection or any acknowledgement of an individual, according to the Gospels, whose adventures were known "throughout all Syria," (Matthew 4:24), or according to Paul (Romans 10:18), whose words reached "unto the ends of the whole world."

Michael Paulkovich offers a far more extensive list of 126 contemporaneous writers, none of whom mentioned Jesus, his exploits or his followers. Among them is the earlier mentioned Philo of Alexandria, the Jewish exegete who wrote in Greek and was well traveled among his contemporaries both geographically and intellectually. Other contemporaries of Jesus who seemed, through their own writings, unaware of his existence included Pontius Pilate, Appian, Cassius Dio, Demosthenens

---

[87] *Treatise on the Gods*, p. 148

[88] *Hellenism and Christianity*, xii

Vaerius Asiaticus, Mamis, Dion of Prusa, Josephus (who lived only a mile from Nazareth yet never heard of either Jesus or Nazareth), Lucanus, Lucian, Pliny the Elder, Pliny the Younger, Plutarch, and Tacitus.[89]

Justus of Tiberias was a Hebrew historian in the latter first century. The Christian scholar Photium states in the ninth century that "Justus makes not the least mention of the appearance of Christ, of what things happened to him, or of the wonderful works that he did." In Matthew (8) it is related that the whole city of Gergesa came out to greet Jesus. Gergesa was a part of Tiberias, yet Justus was apparently unaware of this event.

St. Paul never met Jesus but did experience visions. Yet he writes of the mother of Jesus as an ordinary woman, not a virgin. "Paul knows nothing of Jesus' birth, parents, life events, ministry, miracles, apostles, betrayal, or trial. He knows neither where nor when Jesus lived and died (or ascended). Paul seems to consider the crucifixion metaphorical and spiritual, not physical . . . and he wrote that Jesus was perhaps not a man, but simply a "spirit" of God's son (Galicians, 4:6-14)."[90] Paul claimed to have spent two weeks with the eye witness of Jesus, Peter, and yet knows very little about Jesus. Historians explain the Christian idea of the Trinity as a later addition borrowed from the practices of pagans who began to overwhelm the original Jewish adherents of the faith. The New Testament claim that Jesus grew up in Nazareth has no historical support. Archaeological evidence does not show the presence of an active civilization there until the second century, about the same time that the first New Testament scriptures appeared. The city or village was not even named until several centuries later.[91]

Were first and second century Jews aware of the existence of Jesus? "Our own sources, however, record very little about Jesus' life. Everything that we know about him is found in the Gospels of the New

---

[89] *No Meek Messiah*, p. 199-209. And see his discussion of Paul and his unfamiliarity with any of the details of the life or family of Jesus. According to this source, Paul seemed to believe that Jesus was a spirit and not physical.

[90] Ibid, p. 206

[91] Ibid, p. 215

Testament, a book written by and for the early Christian church."[92] In fact, no contemporary references to Jesus exist in Jewish literature or legal treatises. The first mentions of Jesus appear well after the Jews had been exiled from their land and became subject to the laws of the soon-to-be Christian Roman Empire.

"Both testaments were sadly mauled to make them agree," wrote Menken. "Thus Zechariah XIII, 6, was converted into a prophecy of the crucifixion, Psalms XLI, 9, was made to predict the betrayal of Judas, and Hosea XI, I, came to be accepted as proof of the flight into Egypt, which probably never took place. This laborious tugging and hauling was under way before Paul died, but its best days came afterward."[93]

Remarks by the Jewish sages are extracted from the Talmud and Midrash by Christian apologists in order to support their historical claims. In Tractate Shabbat (of the Babylonian Talmud) 104b, a reference is made to a "Ben Stada" who is identified by one of the sages as "Ben Pandira." This is seen by Christian exegetes as an identification of Jesus because Yeshu ben Pandira was found as a variant of Ben Stada elsewhere in the Talmud. Tractate Sanhedrin 43a remarks that Yeshu was hanged on the eve of Pesach/Passover. And 24 pages later, it says that Ben Stada (who was identified as Ben Pandira) was hanged on the eve of Pesach. All such citations made by Jewish sages are recorded centuries after the existence of Jesus and could be referring to any number of other individuals in a variety of unrelated contexts.

There are commentators that refer to Yeshu, a contemporary of Yehoshua ben Perachya, as Jesus even though he predated the Christian Jesus by a century. It is also known that the term "Ben Stada" was an epithet widely used in Babylonia during the Jewish sojourn there. More significant than the various citations cited by Christian theologians is the fact that those Jewish sages whose words are cited lived long after Jesus was to have lived and preached. None of their predecessors knew of him and their knowledge of his existence would have to have been acquired from the gentile communities that surrounded them. In fact, Pappos ben Yehudah, who the Gemara says was the husband of the mother of Jesus, is the name of a man who lived a century after Jesus,

---

[92] *The Aryeh Kaplan Anthology*, "Why Aren't We Christians," p. 259

[93] *Treatise on the Gods*, p. 229

was a contemporary of Rabbi Akiva, and was said to have been so suspicious of his own wife that he locked her in the house whenever he went out (Gittin 90a).

Authoritative research offers that the passages relating to Jesus "cannot be earlier than the beginning of the fourth century, and is moreover a report of what was said in Babylonia, not Palestine."[94] The attempts to tie the name Pandira through Greek and Aramaic roots to the virginity of his mother are recognized as apologetics contrived along with the much later claim of the virgin birth. And there is no trace of the name Pandira or any near equivalent in the Gospels.

Christian commentators routinely cite the words of Rabbi Eliezer, who lived at the end of the first century, as attesting to the existence of Jesus because he asked about a Ben Stada. If it were actually a reference to Jesus, then it would be the only one attributable to any time period before the compilation of the Gemara. The reference in question is to a passage which describes how R. Eliezer was once arrested on a charge of heresy, which commentators assume was related to Christianity. But there is no evidence that Christianity, as such, even existed during his lifetime.

Many Christian commentaries assume that Talmudic references to Bilaam, the gentile prophet who confronts the Jews in the wilderness after their escape from Egypt, are actually references to Jesus. Some cite the Yalkut Shimoni (thirteenth century midrashic collection)[95] which states that God gave strength to Bilaam's voice so it could be heard all over the world by idol worshipers. In doing so he spoke of seeing "a man, son of a woman, who should rise up and seek to make himself God, and to cause the whole world to go astray." According to the midrash, Bilaam went on to warn the world to not follow that man, referring to a warning voiced in Numbers 23:19 that one should not follow a man that says he is God.

Christian apologists also say that the Gemara (Sanhedrin 43a) admits that Jesus is of royal descent through David. Sanhedrin 107b is also cited as further evidence of the rabbis' knowledge of and rejection of Jesus: "Our Rabbis teach, Ever let the left hand repel and the right

---

[94] *Christianity in Talmud and Midrash*, p. 41

[95] Yalkut Shimoni, Paragraph 766 on the Book of Numbers

hand invite, not like Elisha who repulsed Gehazi with both hands, and not like Rabbi Jehoshua ben Perachya who repulsed Yeshu with both hands." Yet, Yehoshua ben Perachya lived a century before Jesus is thought to have lived. And in Sota 47 there is a reference to Yehoshua ben Perachya fleeing to Egypt to escape Yannai with Yeshu, confirming this earlier dating.

Nazareth does not appear in Jewish writings until about 900 C.E. in hymns. Yet Christian commentators erroneously assume the Hebrew word notzri is a reference to it (through an inaccurate Greek translation-the root of the word is neitzer, meaning sprout or shoot, or may be from natzar meaning to guard or watch). And while the attempt is made to associate notzri with Jesus through its use in the Babylonian Talmud, the Jerusalem Talmud is devoid of such a reference. This is consistent with the chronological understanding that Jewish awareness of Jesus came centuries later, during the compilation of the Babylonian Talmud, and was heavily influenced by the later acceptance of Christianity by the Roman Empire. While the medieval Jewish exegete Rashi refers to Jesus in his commentary on BT Shabbat 104b, his reference (to Egyptian magic) is little more than the promulgation of a firmly entrenched, centuries old Christian legend.

A good summary is offered by Robert Stein. "The key question that arises involves the origin of these rabbinic references. The value would be greatly enhanced if they originated from contemporaries of Jesus who were eyewitnesses of the events they were reporting. . . however, aspects of these accounts seem to be due less to eyewitness reports than to later Jewish interaction with the teachings and claims of the early church." [96]

Finally, the belief that the composition of a special blessing (a 19th) to be added to each of the three times daily Shemona Esrei (18) prayer (also called the Amida) of the Jews might be a nod to the existence of Jesus and Christians is also unlikely. It is believed that Shmuel HaKaton, who lived about 80 years after the Common Era, shortly after the destruction of the second Temple by the Romans, was asked to compose this blessing. This was likely done to reflect and reject the divisions that were manifest among the Jews just before and during the period of

---

[96] *Jesus the Messiah*, p. 34

destruction. There is no evidence that either Shmuel or any of his contemporaries were aware of the existence of either Jesus or Christianity.

Islamic Exceptionalism

"The faith of the Jews prescribed cleaving to the Torah and to the sunna of Moses until Jesus came. Once Jesus came . . . whoever did not reject these and follow Jesus was condemned to perdition. The faith of the Christians consists of adhering to the Evangel and to the laws of Jesus until Muhammad should come. Once Muhammad came, whoever of them did not follow Muhammad … was lost."[97]

The exceptionalist claims of Islam are based on its fundamental belief that the prophecy received and transmitted by Muhammad in the 7th century is the final and perfect revelation to be transmitted to mankind from God. The recipient of that prophecy, Muhammad, is by extension, the greatest of all prophets. Islam must deny the Christian claimant of divinity and emphasizes monotheism as the legacy of Abraham who recognized the One God as a unity. Jesus could not be divine, and because of God's unity could not be part of a tripartite divinity as was the Christian belief beginning in the fourth century. To the Muslim, Jesus was a great prophet, but just a prophet. The Qu'ran teaches the widespread doctrine of Docetism[98] which understands the suffering and death of Jesus to not have been a reality at all. His illusory death is a mere interruption in his career – he remains concealed. He will return from this absence (gaiba in Arabic) to complete his mission. Islam also takes pains to excoriate Judaism for obfuscating the dictates of God to Abraham and ultimately Moses, and corrupting them beyond recognition. Muhammad's role was to eliminate those misunderstandings and restore perfect adherence to faith.

"Islam considers itself as the final plenary revelation in the history of present humanity and believes that there will be no other revelation after it until the end of human history and the coming of the eschatological events described so eloquently in the final chapters of the Qu'ran,

---

[97] *Anti-Judaism*, p. 168

[98] Sura 3, 47-50; 4, 155-156

which is the verbatim Word of God in Islam."[99] It is Islam's task to assert the single religion of the single God in its final form.

The Qu'ran is believed to have been dictated by the angel Gabriel to Muhammad early in the 7th century and then passed on orally to his followers. Muhammad is proudly identified as nabi ummi, an illiterate prophet, and this is a dogma of Islam.[100] Absent an ability to read, he could only have received the knowledge from a divine source. His ignorance helped establish his claims of prophecy and the miraculous origin of the revelation and its Arabic diction. It is believed that the Qu'ran was later assembled in its current form under the supervision of the prophet. Several manuscripts were recorded under the rule of Uthman a few years after the death of Muhammad. The text of the Qu'ran consists of 114 chapters called suras, divided into those revealed to Muhammad when he was in Mecca and those revealed when he later went to Medinah. The order of the chapters is not chronological. The text contains various principles of knowledge, intellect, legalisms, cosmology, psychology, eschatology and metaphysics. The study of the Qu'ran is twofold: tafsir, the study of the outward meaning of the text, and ta'wil referring to its inner meaning. Some of the more mystical inner meaning is taken up by Sufi and to a lesser extent, Shiite tradition.

The Hadith (statements attributed to Muhammad but not appearing in the Qu'ran) are the oral traditions compiled well after the introduction of the Qu'ran. Taken in sum, these extra-scriptural texts are the source of the Islamic legal system (shariah) and early Islamic history. The earliest commentaries on the Qu'ran are based on the Hadith as well. There are different traditions for each of the streams of Islam and each holds some texts more reliable than others. In fact, the historic detail contained there has little or no corroboration. As a result, strict constructionists are increasingly rejecting the authority of the Hadith and relying only on the literal word of the Qu'ran.

The Qu'ran references dozens of the characters and events recorded in the Bible. Muslims argue that the Bible became corrupted (tahrif) over the centuries before its accurate revelation to Muhammad. While

---

[99] *Our Religions*, p. 428-429

[100] *Essential Papers on Messianic Movements,* Israel Friedlander, "Shiitic Influences," p. 136, Also, Sura 7, 156

Islam claims the Qu'ran as an original revelation, its narratives and theme are often taken directly from Biblical accounts and often with purposeful changes and omissions.

The Qu'ran incorporates the origin story of Jesus and Christianity, claiming it as an original revelation from Allah.[101] The text states that Jesus was created in Allah's image just as was Adam. The altered histories of Christianity and Judaism allow the Qu'ran to be very clear about its exceptionalism[102]: "If anyone desires a religion other than Islam (submission to Allah), never will it be accepted of him; and in the hereafter he will be in the ranks of those who have lost all."

The content of the book as a whole is an overt appeal to the Jews, castigating them for their misbehavior.[103] Beginning with verse 40, Sura 2 is a veiled warning to the Jews to avoid the steps of their ancestors who continually sought to subvert the word of Allah and throw away their covenant with God. Reference is made there to the promise of Jesus as well, and Jew and Christian are often lumped together because "Allah will judge between them in their quarrel on the Day of Judgement."[104] But the true believer is warned, "Never will the Jew or Christian be satisfied with you unless you follow their form of religion.[105] The Qu'ran is under the impression that the Jews and Christians study the same scripture, yet disagree.[106] And it borrows and applies as scripture the Talmudic aggada describing how Mt. Sinai was turned upside down and held over the Jews in order to compel them to accept the Torah.[107] The prophet has the Jews admitting to killing Jesus even though they do not realize that the Christian prophet did not really die. Instead, he was elevated to heaven by Allah so that he could testify against the Jews in the future.[108]

---

[101] Suras 35-55

[102] Sura 3:85

[103] Sura 3:69-73

[104] Sura 2:113

[105] Sura 2:120

[106] Sura 2:113

[107] Sura 4:154 and Talmud Bavli, Shabbat 88a

[108] Sura 4:156-159

Who are the Real Chosen People?

Much of Sura 3 is a diatribe against the Jews who would not accede to Muhammad's entreaties after his revelation. The reader is told that the Jews will pay a stiff penalty for their resistance.[109] Among their many indiscretions is the lie that they propagated against Allah by denying themselves non-kosher food.[110] The Jews are the People of the Book, but they rejected the substance of faith and will suffer accordingly. A prelude to that suffering is that of the Jews of Medina and An-Nadir. In the time of Muhammad they were driven from their lands and had their orchards confiscated. When the Jewish community of Banu Quraizah was overrun, 600 men were beheaded and their women and children taken as slaves.

The Qu'ran accuses the Jews of killing their own prophets and as a result, living on in shame. "Soon we shall cast terror into the hearts of the unbelievers . . . their abode will be the fire and evil is the home of the wrong-doers!"[111] "And remember, Allah took a covenant from the People of the Book . . . but they threw it away behind their backs and purchased with it some miserable gain! And vile was the bargain they made!"[112]

The Jews and the pagans are seen to be especially fierce in their opposition to Islam, in contrast to Christians who are neither proud nor arrogant. Christian priests and monks know that the Qu'ran is really true.[113] But, a Muslim must beware that if he befriends a Jew or a Christian, he becomes one of them.[114] It is the responsibility of every Muslim to fight those who reject Islam, including Jews and Christians, until these unbelievers pay regular financial tribute willingly while understanding that they are completely subject to their Muslim overlords.[115] ". . . curses were pronounced on those among the Children of Israel who rejected Faith, by the tongue of David and of Jesus the son

---

[109] Sura 3:86-94

[110] Sura 3:93

[111] Sura 3:151

[112] Sura 3:187

[113] Sura 5:82-86

[114] Sura 5:51

[115] Sura 9:29

of Mary."[116] And, lest one forget, Jews are ". . . the spreaders of war and corruption."[117]

In the Hadith the Jews are described as followers of the Dajjal (what the Muslims call the anti-Christ). When he appears, Dajjal will be accompanied by 70,000 armed Jews. They will be defeated and slaughtered. A variant of that Hadith has Jesus routing Dajjal and his followers in Jerusalem. And on the day of resurrection, the defeated Jews will be consigned to Hellfire, thereby allowing forgiveness to Muslims who have sinned. All Muslim sects quote the Hadith[118] attributed to Muhammad that at the dawn of the messianic era all Jews will suffer violent deaths. The Qu'ran condemns the Jews to perpetual wandering even though they were once given the Land.[119] Consequently, the re-establishment of the Jewish homeland is an affront to Allah and must be reversed.

As if to summarize, al-Tabari (tenth century commentator on the Qu'ran) explains why Christians are somewhat more acceptable to Islam than are Jews: "In my opinion, (the Christians) are not like the Jews who always scheme in order to murder the emissaries and the prophets, and who oppose God in his positive and negative commandments, and who corrupt His scripture which He revealed in His books."[120]

As for common biblical allusions, while the creation accounts of both the Bible and Qu'ran are similar, there are purposeful variations. In taking a harder line than does Judaism in the partnership between God and mankind, the Qu'ran has Allah dictate the names of all living things to Adam who, unlike the angels, is then able to repeat the names. In the Bible, God partners with Adam asking him to name the animals using the genius of the Hebrew language (explored below). The language of the Qu'ran is Arabic, the vernacular of the people to whom it was addressed. As such, Arabic is given no special or holy status. Further, the Qu'ran's creation story identifies Satan as one who is expelled from

---

[116] Sura 5:78

[117] Sura 5:64

[118] Shih Muslim, Book 41, Number 6985

[119] Sura 7:168

[120] Commentary on Sura 5:82, al-Tabari

paradise over his refusal to accept Allah's approval of Adam. It is Satan (Shaitaan) who tempts both Adam and his wife to sin, causing their fall to earth. Contrary to Christian scriptural interpretations, Adam and woman (who with Adam is created from one soul) are forgiven in the Qu'ran -- there is no assignment of original sin.

The story of Cain and Abel is similar in both texts with the exception of the lesson learned from the drama, drawn directly from the Babylonian Talmud (which was in common use by this time)[121], "If one slew a person . . . it would be as if he slew the whole mankind; and if any one saved a life, it would be as if he save the life of the whole mankind."[122]

The deligitimization of Jewish scripture continues in Sura 11 with the story of Noah. Here, changes are designed to be compatible with its necessary divergence from Jewish genealogy. A son of Noah refuses to board the ark and perishes, while some unrelated individuals do board and survive – this allows for a variance from the biblical account that results in the nations of the world spreading out and developing their own tongues. There is no mention of 7 pairs of kosher things, of course, because the idea of kashrut has already been rejected.

Since the Qu'ran is married to the primacy of Ishmael, Abraham's compliance with the command of God to sacrifice his son in Genesis is a story of Ishmael, not Isaac -- who according to Muslim belief was not yet born. And, unlike the Bible, the Qu'ran has Abraham tell his son of God's plan for him.

The story of Lot and the destruction of Sodom and Gomorrah is much the same in both books. But here, in borrowing from the New Testament (2 Peter 2:7, 8), Lot is described as a righteous man, a prophet. The events described in the Bible that follow the destruction of Sodom and Gomorrah and that result in the nations of Moab and Amon are missing in the Qu'ran as neither nation is relevant to the geography or genealogy of Muhammad's spiritual realm.

The narrative of Joseph and his brothers is common to both traditions and differs only in the events that result in Joseph's imprisonment. The Qu'ran mentions the Pharaoh's dream as a single dream while the

---

[121] Mishna, Sanhedrin 4:9; BT Sanhedrin 37a

[122] Sura 5:27-32

Pentateuch sees it as two within one. Events that result in Joseph becoming second in command to Pharaoh and reuniting with his brothers are similar.

A story of Moses is common to both traditions, but in the Qur'an, Moses first attempts to convert Pharaoh to belief in the one God. The advisors of Pharaoh submit to the one God and Pharaoh attempts to fool Moses by feigning belief. Later, in the wilderness, the incident of the golden calf is not facilitated by Aaron as in the Pentateuch but resisted by him. Korach is a rich, arrogant man in the Qu'ran, not the leader of a rebellion against Moses. The roles and narratives of Saul, David, and Goliath are similar. The story of Yonah is altered to leave out Yonah's God given mission to warn the people of Ninevah. Instead it is his own people who are saved by his message. Again, this is consistent with the underlying theme contradicting the Unbelievers and their corruption of Allah's message. Haman is mentioned in the Qu'ran, but he is associated with the Pharaoh of Moses and the building of a tower in Egypt. The Qu'ran also includes the tale of Moses being guided by an unnamed, prophet-like figure (a servant, known to Muslim tradition as Khidir) who reprimands Moses for a lack of understanding due to his spiritual distance from Allah.[123]

In its references to the New Testament, the Qur'an borrows several stories including those of Zachariah and Mary from the Gospel of Luke. Jesus appears in Suras 3, 4, 5, 19, 23, 43, 61 along with other indirect references. These citations give him credit for bringing the dead to life, being the messiah, having disciples, being filled with the holy spirit and being alive in heaven. There are also stories in the Qu'ran about Jesus that are taken from the non-canonical Infancy Gospels.

In addition to the obvious borrowings from the Bible and New Testament and changes to the narrative necessarily implemented to make a case for Islam over both Judaism and Christianity, there is reason to doubt the historicity of the Qu'ran and its origin.

The earliest recorded account of the existence of Muhammad appeared about 40 years after his purported death. The chronicle of the Armenian bishop Sebeos states that Muhammad was a merchant and

---

[123] Sura 18:60-82

that Abraham was a prominent part of his preaching.[124] One hundred years thereafter John of Damascus refers to parts of what is now the Qu'ran, calling it a book for Hagarians, Ishmaelites and Saracens.[125] An inscription attributed to Muawiya, the first Umayyad caliph about 40 years after the death of Muhammad, makes reference to God, but not to Muhammad as a prophet or to the Qu'ran as scripture. Coins from this time period have the name Muhammad on them under the figure of a king bearing the cross of Christianity.[126] The inscriptions on the Dome of the Rock in Jerusalem were completed in 691 CE. They may actually refer to Jesus and to the fact that Muhammad is his messenger.[127] The Hadith were written a century after the death of the prophet. It became necessary at that time to expand the knowledge of the life and behavior of Muhammed as a model for communal observance. There is some evidence that the emirs of the time "forced people to write the hadiths."[128]

While the Qu'ran is known as a work of pure Arabic, some believe that it may have originally been (at least in part) a Syriac Christian liturgical text. Many words do not seem to be of Arabic origin.[129] Some of that language uncertainty is the source for the belief that 70 willing virgins will be encountered in Paradise after the death of a devout Muslim. The language may actually refer to 70 white raisins.[130]

Muhammad never visited Jerusalem and the word does not appear in the Qu'ran. The holiness of that city is a later addition to Muslim tradition. In 638 caliph Omar captured Jerusalem from the Byzantines but prayed facing Mecca as do Muslims in Jerusalem today. An early Islamic name for what was the site of the Jewish Temples was Bauyt al-Maqewdis (Holy Temple). The mosque there is most likely that of

---

[124] *Muhammad*, pp. 73-74

[125] *The Muslim World*, "Of the Tractate of John of Damascus on Islam," John Ernest Merrill, Vol 41, p. 88-89

[126] *Did Muhammad Exist? An Inquiry into Islam's Obscure Origins*, pp. 43-44

[127] Ibid, p. 56-7

[128] Ibid, p. 71

[129] *The Syro-Aramaic Reading of the Koran: A Contribution to the Decoding of the Language of the Koran*

[130] *The Syro-Aramaic Reading of the Koran: A Contribution to the Decoding of the Language of the Koran*, p. 247-283

calif Abd al-Malik who in 688 built the dome of the rock to compete with the Church of the Holy Sepulcher. In fact, the mosque's dimensions are identical to those of the rotunda of that Church.

In Sura 17 of the Qu'ran Muhammad is described as having a dream in which he takes a midnight ride on his flying horse. He is carried from the holy mosque in Mecca to the distant mosque (al-masjid al-Aqsa). The location of al-Aqsa is never mentioned, and is thought by many to be a place in heaven. The text then says that Muhammad ascended to heaven accompanied by the angel Gibril and spoke to both Allah and the prophets.

The Campaign against Jewish Exceptionalism

While the staying power of both Christianity and Islam can be explained by the numbers of their adherents and their political domination, there is little to account for Jewish survival other than psychological and sociological apologetics or a supernatural, divine sponsor. While suffering Judaism's continued existence, both Christianity and Islam embrace a bedrock belief in Judaism's ultimate demise. Yet the Jew persists.

Maimonides summarized the religious oppression of the Middle Ages: "Can there be a greater stumbling block than (Christianity)? All the prophets spoke of Moshiach (the messiah) as the redeemer of Israel and their savior, who would gather their dispersed ones and strengthen their (observances of) the mitzvahs. In contrast (the founder of Christianity) caused the Jews to be slain by the sword, their remnants to be scattered and humiliated, the Torah to be altered, and the majority of the world to err and serve a god other than the Lord . . . all the deeds of Jesus of Nazareth and that Ishmaelite who arose after him will only serve to pave the way for the coming of Moshiach and for the improvement of the entire world, (motivating the nations) to serve God together, as it is written (Zephaniah 3:9), 'I will make the peoples pure of speech so that they will all call upon the Name of God and serve Him with one purpose . . . When the true Messiah king will arise and prove successful, his (position becoming) exalted and uplifted, they will all return and realize that their ancestors endowed them with a false heritage, their prophets and ancestors cause them to err."[131]

ALTERNATIVE FACTS AND CHRISTIAN SUPPRESSION

According to the modern intelligentsia there are myriad reasons for Jew-hatred, none of which finds its source to be the religion itself. H.L. Menken elaborates: "As commonly encountered, they (the Jews) strike other peoples as predominantly unpleasant, and everywhere on earth they seem to be disliked. This dislike, despite their own belief to the contrary, has nothing to do with their religion: it is founded, rather, on their bad manners, their curious lack of tact. They have an extraordinary capacity for offending and alarming the Goyim, and not infrequently, for the earliest days down to our own time, it has engendered brutal wars upon them. Yet these same rude, unpopular and often unintelligent folk, from time immemorial, have been the chief dreamers of the Western world and beyond all comparison its greatest poets ... All this, of course, may prove either one of two things: that the Jews, in their heyday, were actually superior to all the great peoples who disdained them, or that poetry is only an inferior art."[132]

The Babylonian Talmud spends page after page discussing the legal intricacies of many a seemingly small topic, but in tractate Shabbat, page 89a, it devotes just three lines to the topic of anti-Semitism. "What is the reason for the name Har Sinai (Mt. Sinai)? That hatred (sina) descended to the idolaters on it." To the sages it was a simple fact that the root of anti-Semitism was the acceptance of the Torah by the Jews. Jew "hatred" is as integral a part of nature as the laws of physics and there is nothing that can be done to change it. And, it has spawned inter-ethnic reactions that are most often justified by the language of social science, not religion.

Attacks on Judaism are ad hominem in nature, unlike those levied on any culture or religious belief extant or extinct.

Karl Marx believed that money is the God of Judaism. As long as there is private property, Jews cannot be emancipated, nor can society

---

[131] Rambam on the Messiah; all texts after 1574 were censored.

[132] *Treatise on the Gods*, p. 286-7

be emancipated from the Jews. In his *Against Apion*, Josephus retells Apion's introduction of the idea of the Blood Libel as an example of Jewish hostility to all aliens. Even something so neutral and seemingly apolitical as *The Oxford English Dictionary* lists one definition of the proper noun Pharisee as "hypocrite."

Medieval Christians came to believe the "revenge of the savior" story. They believed that Roman emperor Vespasian was a divine tool used to conquer the Jews and destroy Jerusalem as punishment for Jewish deicide. While this became something of a model for the relationship between a Christian monarch and his Jewish subjects it was also reason for him to keep a remnant of the Jews alive and under his protection. This was a logical extension of the position taken by St. Augustine who insisted that the preservation of the Jews was essential because of their value as witnesses to their own exile.

The Protestant Reformation did nothing to retard Christian anti-Semitism. According to Martin Luther, "They (the Jews) are given to all people in the whole world to tread down, just like scum in an alley, which is thrown out because it is of absolutely no use to anyone, except to soil one's feet . . . just like anything of worth is excluded from dust and scum, so also from the castaways, so that there is nothing left among them that can still be of use to anyone . . . One must beware of the shit of Rabbis, who have in a sense made the Holy Scripture into a latrine of sorts, into which they may introduce their shameful pursuits and utterly stupid opinions."[133]

The philosophers of the Christian world were no less kind. Kant believed that Christianity owed nothing to Judaism. He placed Judaism below all other philosophies of religion and, strictly speaking, did not see it as a religion at all. Unlike the other religions, which in varying degrees might ultimately develop into the pure religion of reason, the "Jewish religion stood out for Kant as having no moral content at all; it was merely legalistic, a political constitution only."[134]

Even the literary examination of the Jewish Bible in the 19th century was suffused with an embedded Christian anti-Semitism. The methodology for the literary analysis and exploration of the Pentateuch,

---

[133] *Anti-Judaism*, p. 255

[134] *Dark Riddle*, p. 7

developed by Julius Wellhausen,[135] came to be called the Documentary Hypothesis. The originator of this school of higher criticism could not resist the urge to declare that the Jewish Bible "blocks up the access to heaven . . . and spoils morality."[136]

By the time of the Second World War, "The Nazis had a comparatively easy time of it. There was great understanding evinced for their anti-Semitism the world over. . . To what extent demoralization had engulfed the West may be gauged by comparing the attitudes of successive American governments toward pogroms and Jewish persecutions in Russia, Rumania, Turkey in the nineteenth and early twentieth century, with the forbearance toward Nazi Germany."[137]

In a series of polls taken by the Opinion Research Corporation between 1939 and 1946, Americans were asked to name the greatest threat to the United States and consistently chose the Jews over the military threats of the Japanese or the Germans. Their fear peaked in June of 1944, just as the Jewish population of Europe was nearing extermination.[138]

Eliezer Berkovits concluded, "There is sufficient evidence to believe that an ingrained theological anti-Semitism that for long centuries had nourished Christianity and was responsible for a tradition of Christian inhumanity toward the Jew was ultimately responsible for the spiritual madness of encouraging acquiescence."[139] "As a result of Christian theology, teaching and tradition, the feeling among Christians was widespread that the Jews were receiving what was due them." And there was no attempt to hide such feelings. After escaping the Nazis and making his way to the Papal Nuncio of Slovakia in order to describe the imminent deaths of thousands of his fellow Jews, escapee M.D. Weissmandel was told, "There is no innocent blood of Jewish children in the world.

---

[135] *Prolegomena to the History of Israel*, Julius Wellhausen, Charleston, BiblioBazaar, 2007.

[136] *Anti-Judaism*, p. 455

[137] *Faith After the Holocaust*, p. 12

[138] *Anti-Judaism*, p. 458

[139] *Faith After the Holocaust*, p. 16

All Jewish blood is guilty. You have to die. This is the punishment that has been awaiting you because of that sin."[140]

Father Flannery wrote that the Pauline doctrine is the true expression of Church teaching. Jews are "enemies for your sakes," "the branches that were broken off because of unbelief;" they are the faithless sinners, but "they are beloved for their father's sake."[141]

Giving no credence to the Talmudic explanation, sociologists and proponents of replacement theologies would like to believe that "Anti-Semitism is not, despite a common opinion, as old as the Jews. While occupying a homeland of their own, Jews encountered the normal hostility of rival powers but nothing that could strictly be called anti-Semitism. This development was reserved for the Diaspora, the dispersion, and it was not until the third century B.C.E. that its presence there could be clearly discerned." As one Greek historian put it, ". . . in remembrance of the exile of his people (Moses) instituted for them a misanthropic and inhospitable way of life."[142]

Christian anti-Semitism accelerated in third century Rome. Shortly thereafter, the synod of Elvira in the year 306 prohibited the intermarriage and even the sharing of meals between Christians and Jews. During the reign of Justinian (527-564) conversion from Christianity to Judaism was prohibited. In the several centuries that followed, Christians were prohibited from being treated by Jewish doctors and from living in Jewish homes. Taxes were levied forcing Jews to support the Church, special clothing was required for Jews, and Jews were prohibited from attending universities but were required to attend church sermons.

The idea that the Jews were responsible for the death of the Christian savior was first voiced by St. Justin. The opportunities to visit both revenge and punishment on the perpetrators came often but reached full flower in the Christian Middle Ages. The first Crusade (1096) left thousands of dead Jews in its wake and became the prototype for successive "mobile" persecutions. The first recorded European blood libel took place in 1141 in Norwich and was duplicated in every other European

---

[140] *Faith After the Holocaust*, p. 17

[141] Ibid, p. 25

[142] *Anguish*, p. 7 & p. 12: Hecataeous of Abdera, Greek historian, early third century BCE, Reinach, no. 9, p.17, *Anguish*, p.12

country. The Black Death of the 14th century was blamed on the Jews, resulting in many being burned alive or hanged by mobs.

Beginning with an anti-Jewish riot in Odessa in 1821, the pogrom became a popular instrument of Christian terror and suppression. From 1881-1884 pogroms swept through the Jewish communities of the Ukraine and southern Russia following the assassination of Czar Alexander II. Belarus and western Ukraine saw much of the same during the civil war that followed the 1917 revolution, killing tens of thousands of Jews. And none of this includes the persecutions that preceded the Spanish Inquisition, the expulsion from every Christian country in Europe and, later, the Holocaust.

Even where Jews did not exist, intellectuals and political leaders pointed to their guilt. Pascal saw the misery of the "carnal" Jew as proof of Christianity.[143] Late in the 19th century, Christian convert, Jacob Brafman propagated the charge of a world-wide Jewish conspiracy, which in 1919 became the basis of the *Protocols of the Learned Elders of Zion*.[144] In Stalinist Russia, Jews were again singled out. "It did not take long to see that anti-Zionism meant little more than anti-Semitism: that Jews of every hue and strain, even anti-Zionists, were the targets."[145]

Johann Andreas Eisenmenger, a 17th century contemporary of Spinoza,[146] studied rabbinic literature and was a student of rabbinic leaders in Amsterdam and Palestine. He later turned against Judaism, writing *Entdecktes Judenthum* (Judaism Unmasked or Revealed). In it he collected citations from 193 books and rabbinical tracts in order to attack Judaism. Many of his sources were the work of Jewish converts to Christianity. "Eisenmenger proceeded to amass quotations from the Talmud and other Hebrew sources revealing to all how the Jewish religion was barbarous, superstitious, and even murderous. All this was done in an apparently scholarly and reasonable way that belied the

---

[143] Ibid, p. 149

[144] Ibid, p. 173

[145] Ibid, p. 238

[146] Spinoza's biographer, Jean Maximilien Lucas praised him as "being born and bred in the midst of a gross people who are the source of superstition, he has imbibed no bitterness whatever, and . . . he was entirely cured of those silly and ridiculous opinions which the Jews have of God." As quoted by David Nirenberg, *Anti-Judaism*, p. 339

author's evident preoccupation (like Luther) with tales of Jewish ritual murder of Christian children and poisoning of wells ... Eisenmenger ... recommended abolishing their (the Jews) present freedom in trade ... making them lords over the Germans. He demanded too an immediate ban on their synagogues, public worship and communal leaders and rabbis."[147] The book was initially suppressed at the request of German Jews who attempted to bribe Eisenmenger. Ultimately it was widely disseminated and respected for its source material.

Today, with the restoration of a Jewish presence in the State of Israel, anti-Semitism finds a political banner under which to hide: "(anti-Zionism) has proved, moreover, to be an incredible unifier, often bringing Arabs, Communists, Protestants and Catholics, Leftists and Rightists, Blacks and Whites under the same banner, thus demonstrating the same virtuosity that the age-old anti-Semitism it disclaims enjoyed."[148] Father Flannery concludes, "A ubiquitous myth has taken root: an alien people (the Jews) have expelled an indigenous people (the Arabs) from their homes, forcing them to fester in poverty on the borders of their own homeland. The myth, incredibly, has been swallowed whole by many otherwise intelligent and fair-minded people without the least effort to verify any part of it."[149]

ALTERNATIVE FACTS AND MUSLIM OPPRESSION

Arab literature was replete with praise for the Jewish people before the advent of Islam.[150] It is said that when he began his rule in Medina, Muhammad adopted both prayer facing Jerusalem and the fast of Yom Kippur in order to solicit prospective Jewish followers. His nearly universal rejection fed the need in the emergent Islamic world to suppress and persecute non-believers. According to tradition, Muhammad said on his death bed: "May God fight the Jews and the Christians! They transformed the tombs of their prophets into mosques. Two religions

---

[147] *Revolutionary Antisemitism in Germany: From Kant to Wagner*, pp. 8f

[148] *Anguish*, p. 268

[149] Ibid, p. 270

[150] *The Arabs in History*, p. 31-32

will not remain in the land of the Arabs."[151] The Qu'ran codifies this malevolence toward the Jews: "Their effort is for mischief in the land, and God does not love mischief makers."[152]

After turning away Muhammad, Jews were expelled from Medina, hundreds being murdered. They were thereafter forced to pay a tax to their Muslim overlords. In addition, they were prohibited from public office, armed service, camel or horse riding, synagogue building, public consumption of wine, loud prayer, and the presentation of legal evidence against a Muslim. They were forced to wear distinctive clothing as well.

In the 8th century, the Muslim ruler Idris I destroyed several Jewish communities. The Almohads did the same in North Africa in the 12th century. In 1066 the Arabs of Grenada razed the Jewish quarter of the city and slaughtered its 5000 inhabitants. In 1465 the Arabs of Fez slaughtered all but 11 Jews in the city and a series of massacres throughout Morocco ensued. In the late 19th century hundreds were murdered in Marrakesh and earlier, in 1785 Ali Burzi Pasha murdered hundreds. Massacres took place in Algeria in 1805, 1815 and 1830. Jews were forced to convert or die in Yemen in the 12th and 17th centuries, in Morocco in the 13th, 15th and 18th centuries, and in Iraq in the 14th century. Jews were isolated in ghettos in Algeria, Tunisia, Egypt, Libya and Morocco in the 19th century. Blood libels against the Jews became routine in Muslim lands.[153]

Muslim poet, Ibn Hazm (11th century Spain) believed that the historic role of Jews was to bribe Paul to spread false teachings and trick the early Christians into believing that Jesus was divine instead of being just a prophet. He also insisted that the Jews were responsible for the spread of schism within Islam from its inception. As if presaging the later words of the medieval church, Martin Luther and Karl Marx, Ibn Hazm insisted that Jews were driven by their materialism. "The religion of the Jews tends strongly towards that, for there is not in their Torah any mention of the next world, or of reward after death."[154]

---

[151] *Anti-Judaism*, p. 163

[152] Sura 5:64

[153] *The Jews of Islam*, p. 158

[154] *Anti-Judasim*, p. 181

In so far as the Qu'ran's literary treatment of the Jews is mirrored by actual experience, Maimonides summarizes[155]: "Never did a nation molest, degrade, debase and hate us as much as they . . . although we were dishonored by them beyond human endurance and had to put up with their fabrications . . . We have acquiesced, both old and young, to inure ourselves to humiliation . . . All this notwithstanding, we do not escape this continued maltreatment (by Muslims) which well nigh crushes us. No matter how much we suffer and elect to remain at peace with them they stir up strife and sedition."

Modern rejection of the Jew is fundamental to the Islamic mindset. "Our hatred for the Jews dates from God's condemnation of them for their persecution and rejection of Isa/Jesus and their subsequent rejection of His chosen prophet . . . for a Muslim to kill a Jew, or for him to be killed by a Jew ensures him an immediate entry into Heaven and into the august presence of God Almighty."[156] Contemporary Islam does nothing to discourage an altered historical narrative that asserts as fiction the existence of the Jewish Temples and as fact modern Palestinian Arabs as indigenous Jerusalemites. As a result, modern Islam is actively engaged in the manufacture of human weapons, devoid of historical knowledge who kill to honor an imaginary past. Claiming God's favor as His chosen has created an irreconcilable conflict which is not cultural but religious.

Christian and Muslim engagement with the Jewish community is part of a modern, Western, sociological interreligious belief system. We are all the children of God, are we not? Yet, the fact remains that Jews have been singled out for scorn in the origin stories of each of these competing religions. The sincere and sympathetic effort to subvert the original, scriptural Christian and Islamic demonization of Judaism is now overwhelmed by the reality of a thriving State of Israel. Historically tolerated for millennia within the confines of the Christian and Muslim diaspora, Jews are now the face of their own state, sparking a violent re-emergence of fundamental theology.

---

[155] Maimonides' *Epistle to the Jews of Yemen*

[156] Foreign Office File #371/20822 E 7201/22/31; Elie Kedourie, Islam in the Modern World, London, Mansell, 1980, p. 69-72 quoting Saudi King Ibn Saud speaking to British Colonel H.R.IP. Dickson on November 23, 1937.

If Judaism and its outgrowths are to co-exist peacefully, then each must examine its origins in place, necessity, sociology and history. Christianity and Islam must confront their late and damaging claims to "choseness" and instead invest intellectual capital in the enrichment of their own devotional and social behaviors, abandoning proselytization and marginalization of outsiders. Judaism must do the same and in the process embrace its role in a God-fearing religious network.

Is antisemitism a product of sociology or divine planning? Can the humanist explain why one "ethnic" group has been singled out for scorn and mistreatment during millennia of nation building? Can he explain why the Jew continues to be held to a standard different from that of any other in this, the era of enlightenment and democracy? The world's focus on the Jews and its routine condemnation of their ethnic and national behavior is so wildly out of proportion to their presence as to be absurd. It is an historical fact that the nations of the world segregated the Jews from themselves until at least the Enlightenment. For the last three centuries Jews have worked tirelessly to integrate themselves into Western society, their greatest hope being societal equality. They are celebrated for the value that they bring to host societies, much as they were in the medieval Muslim world. But the Jewish physician to Muslim royalty or financier to European kings knew he was hated and had no hope of social acceptance. Today, Jewish physicians, scientists, financiers, lawyers, academics and authors are often welcomed by their gentile peers. But acceptance is gained at the cost of a diminished or rejected heritage. In the mind of the modern Jew, the idea of "chosen" status is antithetical to the quest for acceptance. If the goal of the Jewish people is equality with their gentile peers, then chosen status is an unaffordable encumbrance.

V

MESORA AND THE LIMITATIONS OF JEWISH LEADERSHIP

WHAT HATH HISTORY WROUGHT!

The Pentateuch -- the writings attributed to Moses as dictated or inspired by God -- is a diatribe against pagan lifestyle and ideology. It demands, often in dramatic and terrifying language, the rejection of practices that smack of image worship and the denigration of human life. So essential was it that pagan habits be broken that the wrath of God is frequent and directed at those who would influence His designated people toward such practice.

According to scripture, the Jews were formally transformed from a family into a nation during their 210 year sojourn in Egypt, the last 86 years of which became enslavement to their hosts. Under God's direction, Moses led 600,000 men, their families and allies out of Egypt (as many as 4 million people according to many) into the wilderness and to the land that God had promised to their forefathers Abraham, Isaac and Jacob.

Forty years later, after the death of Moses, his successor Joshua began the conquest of that land. With his passing a series of judges led the tribally divided nation until several generations thereafter, under the supervision of the prophet and judge Samuel, a king was selected.

David, the second monarch to rule, began both the process of establishing Jerusalem as the nation's capital city and locating the Holy Temple there.

With the death of David's son Solomon, a dispute over royal succession split the kingdom into two, and within a few generations the Northern Kingdom (also known as Israel) disappeared into the hands of the Assyrians. In Judah (the Southern Kingdom), the Temple and its priests remained the center of religious life while the king was the undisputed civil monarch. In the sixth century before the Common Era Jerusalem was destroyed by the Babylonians and much of the population was carried off into captivity. Within a century only a small portion of those Jews returned to begin rebuilding the Temple.

Late in the Second Temple period, about a century and a half before the Common Era, the traditional leadership of the Jewish nation broke down as the Jews successfully fought off the Hellenizing Greeks. Yet, while the Jewish commonwealth was saved, political and religious leadership was co-opted and consolidated by a single, priestly family – the Hasmoneans. The unification of religious and civil leadership resulted in the gradual decimation of trust in both. This new reality spawned resistance movements within the religious community.

At the same time, the Jewish diaspora grew beyond Persia throughout the Mediterranean.

"Considerable light has been thrown on the kaleidoscopic variety of Jewish messianic beliefs during the end of the Second Temple Period by the discovery in the Judean Desert of the library of the congregation of the New Covenant (also known as the Dead Sea Sect). The sect is remarkable among other things for envisaging a future presided over by two anointed heads: a Davidic king and (superior to him) a messianic High Priest of the House of Aaron. It seems obvious that here the messianic pair functions as a sign and symbol of the perfect social order . . . "[157]

Power, wealth and (ironically) the Hellenization of society became the primary focus of this new, unified leadership model. Jewish society and practice was permanently fractured. By the time of the reign

---

[157] *Messianism in Jewish History, Essential Papers on Messianic Movements and Personalities in Jewish History*, R.J. Zwi Werblowsky, p. 41

of Herod, onerous taxes had been imposed to support venal building projects and suffocating security measures. The country was essentially bankrupt.[158] New pockets of leadership arose at the grass roots, while a rabbinic cult provided religious prescriptions for daily life and at the same time rationalized and continued Temple practice. Elements of the oral law that would dictate future Jewish religious practice took shape as the boundaries of personal behavior became rabbinically defined. The influence of the rabbinate expanded, particularly outside of Jerusalem. From this dissonance arose a diverse offering of religious options, one of which became the cult of Jesus.

Several socio-religious choices gained currency. Many Jews simply assimilated into Greek society – the very action that led to the Hasmonean revolt. Others adopted the Greek language but were still loyal to their Jewish tradition. They read the Septuagint (the Greek translation of the Torah) and later followed Philo of Alexandria and other Greek speaking biblical exegetes. According to Josephus, the Sadducees and Boethusians attracted adherents based on their belief in free will and their denial of individual providence. First identified by name about a century and a half before the Common Era, the Pharisees also rejected Hellenistic culture. They adhered to what they believed were ancient religious traditions and focused most of their attention on the rural, lower socio-economic class. The earlier mentioned Dead Sea sect rejected both Greek influence and language. Many were members of the priestly class and had been expelled by the Hasmoneans. Their writings demonstrate a selective rejection of many rabbinic dictates.

By the end of the Second Temple period these groups were actively squabbling over jurisdiction in Jerusalem. Riots focusing on the conduct of the priestly sacrificial service took place in both the Temple itself and its surroundings, and were alternately led by both the Sadducees and Pharisees. There was also a "fourth philosophy" group (according to Josephus) that revolted against the Romans from 4 B.C.E. onward with a pharisaic philosophy (Sicarii). Because the majority of the population was less well-educated, less affluent and lived outside of

---

[158] *Antiquities*,17.11,2 para 307-8

Jerusalem it became the focus of the rabbinic cult (the Pharisees) after the destruction.[159]

With the failure of the Bar Kochba revolt against Roman control several decades later, and an inability to continue to define Judaism in terms of its relationship to the Holy Land, the race to find alternatives to rabbinic leadership accelerated. But these competing theological approaches were not addressed in the writings of the rabbinic cult. In fact, the Mishna was assembled in its final form early in the third century with, at most, minimal influence from or recognition of any specific breakaway theologies. By the time of the Jerusalem Talmud and the later Babylonian Talmud, both those influences and the acceptance of Christianity by the Roman Empire were firmly in place.

SURVIVAL MODE

While it seems counterintuitive, after the destruction of Jerusalem and the dispersion of most of its Jews, the practice of Judaism and the development of its institutions evolved *toward* Christianity and Islam rather than away from them. Torah law and life had to be developed and maintained in the face of radical social and political change – and that had to be done under the watchful eyes of Gentile hosts.

Consequently, daily Jewish life bears the legacy of 2000 years of displacement, rejection, persecution and slaughter. As one would expect, contemporary, traditional observances reflect that experience.

Take for example this description of the contemporary observance of Tisha B'Av, the ninth day of the Hebrew month of Av, the day that commemorates the destruction of both ancient Jewish commonwealths:

> Dozens are crowded into the social hall, just behind the synagogue sanctuary. Chairs have been cleared away, neatly stacked eight-high in the corners. Friends and neighbors have arrived carrying step stools or milk crates. The only other seating alternative is a hard and unforgiving tile floor. Some early arrivals have propped themselves up against an outside wall. The lights are

---

[159] *Challenge and Transformation: SecondTemple and Rabbinic Judaism*

dim and no greetings are offered. The evening service has already been said and a reader is about to begin the recitation of the biblical book of Eicha (Lamentations). Credited to the prophet Jeremiah, the text is poetic testimony, describing the suffering experienced during the destruction of the Temple in Jerusalem at the hands of the Babylonians more than two and a half millennia ago.

The annual observance of this calamity is fraught with signs of mourning. Discomfort is obligatory. Fasting and abstinence from signs of pleasure like and washing and wearing leather shoes are imposed. The ninth day of the Hebrew month of Av falls out in the hottest month of the summer in the Northern Hemisphere. It commemorates not just one, but a number of disasters that have befallen the Jewish people. Tradition traces the day back to the biblically recorded exodus of the Jews from Egyptian slavery, when in the second year God directed the people to enter the land that He had promised to their forefathers. But the fear, reluctance and doubt displayed by the advance scouts at that time discouraged the masses. Their recalcitrance resulted in an additional 38 years of wandering in the desert, during which time an entire generation expired. The destruction of both temples in Jerusalem, the Jewish expulsion from Spain, the beginning of WWI and a number of other tragedies are said to have occurred on that day. The first nine days of the month of Av are observed with caution, privation and care.

"How could you hasten your wrath, ruining your loyal people at the hand of Rome and not remember your covenant with Abraham?"[160] A stubborn people, Jews have often been their own worst enemies. Yet, they demand that God intervene and rescue them from their misery. From the start this chosen people rebelled even after witnessing God's

---

[160] From a poem written by Elazar Hakalir, one of many read on the day of Tisha B'Av

miracles. And that behavior became all the more ingrained with the passage of time and the acceptance of suffering. Much of this was foreseen by the Torah itself. Jews were destined to lose faith and invite punishment. Scripture predicted that they would be the smallest of peoples and that other nations would fear and resist them. But they were also given a formula for covenantal redemption and success. For almost 1500 years, the Torah narrative played itself out as the Jews were enslaved, freed, and eventually settled in the land promised to them. Their efforts to establish a kingdom of priests saw some early success, but later succumbed to the biblically forecasted diversions and distractions that led to the predicted punishments. It seemed historically inevitable that at some point a preoccupation with survival would take precedence over the chosen role.

The Jewish people had been chosen by an all-powerful God and Creator as His possession. Early in this nation's existence, "exceptional" status was integral to its conquest of the land, survival in exile and eventual return. However, the Hellenization process of the centuries preceding the Common Era -- the destruction of the divine leadership model by the Hasmoneans who combined the roles of priest and monarch -- and the subsequent assumption of both civil and religious authority by the Pharisees – the predecessors of the rabbinic cult -- began the process of reducing Jewish exceptionalism to an historic footnote. Its fate was sealed with the failure of the rebellion by Palestinian Jews against Rome in 132 C.E. the intention of which was the restoration of the divinely promised commonwealth. Jews were dispersed among the nations and took direction from their rabbinic leaders to live among the gentiles, abiding by their laws in a quest for survival until the ultimate redemption.

In the process of trying to provide a spiritual and practical home for its displaced people, the rabbinate inadvertently abetted the contemporary emasculation of ancient Jewish exceptionalism. Their effort to rescue the biblically chosen people included the imposition of increasingly restrictive religious laws that both undercut the "chosen" designation and encouraged secular assimilation. Jewish religious observance stood in the shadow of the Mishna's command to build a "fence" around the

Torah.[161] Strict adherence to rabbinic law was primary as observance took on an increasingly narrow character. The mandates of the oral law became an unintentional but unrelenting burden, imposing a strict definition of fealty on a constituency deemed increasingly incapable of making responsible religious and secular judgments. Originally created to guard against the violation of Torah intent, these growing restrictions instead signaled an institutional lack of faith in the efficacy and currency of ancient covenantal promises.

Contemporary Judaism now tolerates an uncomfortable co-existence with Christianity and Islam as an older, but less significant partner vying for acceptance by both God and society. Judaism's "Mesora" (tradition) is at once the glue that binds historically and geographically disparate Jews together, while it simultaneously obscures the divinely envisioned chosen role. As a result, it is not just science and the "upstart" Christian and Muslim supersessionists who unknowingly conspire to derail the divine plan. God's designated "chosen" people are complicit as well.

The Mesora as a Weapon

The modern state of Israel is both the political and social surrogate for a visceral antisemitism. European Jews are under constant physical threat. American college campuses are fertile ground for both physical and psychological intimidation. The United Nations promulgates an alternative historical Middle Eastern narrative. Public manifestations of hatred are increasingly common. The Boston affiliate of National Public Radio chose to shut down its web site after the response to its 8 minute report on the rise of internet anti-Semitism[162] elicited a flood of vividly descriptive racist, anti-Semitic comments from listeners. In the world's most tolerant democracy, little has changed over three generations. And with a disruptive political climate in Europe and an influx of Muslim refugees from the Syrian and Lybian civil wars, the pushing

---

[161] Avot, 1:2

[162] *Algemeiner*, 6/15/16, "NPR Producer: Deluge of Jew-Hatred in Response to Broadcast About Online Antisemitism."

of a Jewish doctor to her death from her apartment window in France raised nary an eyebrow.

Contrary to accepted psychological rationale, this historic, societal distaste for the Jewish people has not united Jews, but has instead been internalized. It manifests itself in the form of distrust, exclusion and competition between practicing Jewish communities. Campaigns to discredit one portion of the community at the expense of another are all too common. Contemporary religious leadership seems content to fracture communities rather than embrace diversity.

In tractate Taanit, pages 20a/b, the Babylonian Talmud addresses the arrogance of Torah leadership and its predilection to sometimes use learning as a tool to denigrate others.[163] "Our rabbis have taught: A man should always be gentle as the reed and never unyielding as the cedar. Once Rabbi Elazar the son of Rabbi Shimon was coming from Migdal Gedor from the house of his teacher, and he was riding leisurely on his donkey by the riverside and was feeling happy and elated with himself because he had studied a great deal of Torah. By chance he met an exceedingly ugly man who greeted him, 'Peace be with you, sir.' Rabbi Elazar did not return his greeting but instead said to him, 'Raicha (empty one, someone with no redeeming value). How ugly you are. Are all the citizens of your city as ugly as you are?' The man replied, 'I do not know, but go and tell the craftsman who made me, 'How ugly is the vessel which you have made!' When Rabbi Elazar realized that he had sinned he dismounted from the donkey and prostrated himself before the man and said to him, 'I ask you to please forgive me.' The man replied, 'I will not forgive you until you go to the craftsman who made me and say to him, 'How ugly is the vessel which you have made.' The rabbi then walked behind him until he reached his city. When his fellow citizens came out to meet him they greeted him with the words, 'Peace be upon you Our Teacher, Our Master.' The man asked them, 'Who are you addressing in this way?' They replied, 'The man who is walking behind you.' He then exclaimed, 'If this man is a teacher, there should be no more like him in Israel.' The people then asked him why. He replied, 'Such and such a thing has he done to me.' They said to him, 'Nevertheless, forgive him, for he is a man of great Torah learning.' The

---

[163] My thanks to Kayla Avraham for pointing this out.

man replied, 'For your sakes I will forgive him, but only on the condition that he does not act in the same manner in the future.' Soon thereafter Rabbi Elazar entered the study hall and expounded that a man should always be gentle as the reed and never as unyielding as the cedar. For this reason the reed merited that of it should be made a pen for writing the law, tefillin and mezuzot."

This lesson appears to be lost on contemporary religious leaders who have fallen into the habit of repeating similarly divisive behavior. In a period of just over a month in a handful of large American Jewish communities, an orthodox rabbi obsessed over whether to permit an openly gay congregant to pray in his synagogue; a local colleague of his routinely reviewed the email posted on a community bulletin board in order to censor messages that might promote speakers or community activities not in keeping with his religious ideology; clergy at a local kollel (an institution for the advanced learning of Torah) and at that city's orthodox rabbinical organization were instructed to "freeze out" a more liberal congregation so as to throw its legitimacy into question; a long tenured congregational rabbi in a major metropolitan area held a series of classes for the community on the evils of what has become known as Open Orthodoxy; a scion of the country's leading modern orthodox educational institution wrote that he regretted the success of contemporary Torah education for women, and in particular the program initiated some years ago by the university that is his employer; and, a rabbi and teacher at a well-known yeshiva (secondary and post-graduate institute for Torah education) asked the creator of a popular Jewish rock band, one that uses its music to teach religious values, to cancel its performance at a local orthodox synagogue (whose practices he disagrees with) so that fans would not "lose faith in you and your holy mission."

These destructive actions were taken by those who feel that they are "duty-bound" to defend the "eternal mesora (traditional practice) community of authentic Orthodoxy."[164] In fact, this behavior thinly veils the profound insecurity evidenced by that community, encourages cynicism among its constituents, and breaks down respect for the institutional leadership of communal religious organizations in general. Representatives of Jewish orthodoxy are defending a tradition said to

---

[164] TorahWeb.org, Rabbi Mordechai Willig, "Trampled Laws," August 7, 2015

date back to Moses at Sinai. And while the historical validity of that belief is at best questionable, its defenders insist that its authority supersedes that of the written Torah and even the foundations of the oral law codified in the Mishna.

Contemporary sages are open in their belief that, "we do not need realism (historical truth), we need inspiration from our forefathers in order to pass it on to posterity."[165] Nineteenth century Rabbi Samson Rafael Hirsch said about our great Torah icons and sages, "They do not require our apologies, nor do such attempts become them. Truth is the seal of our Torah, and truthfulness is the principle of all its true and great commentators and teachers."[166] Should those not well versed in religious law and practice be deliberately misled in order to assure fealty to the existing model of religious leadership – a model close to that employed by both Christianity and Islam?

"The Pharisees, exponents of the principle that the Roman state exercised legitimate authority so long as religious law was not violated, made every effort to restrain the people from revolting against Rome."[167] The new rabbinic class subjugated any political aspirations to daily religious practice. A rabbinic mesora was launched that seemed to betray a growing uncertainty in the fulfillment of God's covenantal promise. Without the guidance of law and its associated mesora, so the reasoning went, Jews would disappear into the fabric of surrounding societies. Without rabbinic efforts to keep the Oral Torah alive in a fixed and public form it would surely disappear.

The battle over Jerusalem around the destruction of the Second Temple was a battle over both political primacy and faith in the scriptural and oral traditions.[168] Upheaval, dispersion, assimilation and the decimation of faith followed. Was the sacrifice of Jerusalem by the rabbinic leadership the wisest choice?

---

[165] Rabbi Shimon Schwab quoted by Marc Shapiro on page 3 of *Changing the Immutable*

[166] R.S. Hirsch, Commentary on the Pentateuch, Genesis, Chapter 12, verses 12,13

[167] *A Hidden Revolution*, p. 60

[168] Not to be confused with the contest over acceptance of the oral law between the Sadducees and Pharisees

The Babylonian Talmud actually touches on this topic. It describes the contrarian view of Rabbi Akiva (Gittin 56b). He argues that Rabbi Yochanan ben Zakkai (at the moment of Jerusalem's destruction) erred by only asking that the Roman Emperor Vespasian allow his safe passage from Jerusalem to Yavne where the rabbinic tradition would be secured. Rather, in the eyes of Rabbi Akiva he should have asked that Jerusalem be spared entirely. Rome's vigilant retreat may have allowed the Jews to resolve their problems internally. The later defeat of Bar Kochba at the hand of Rome, despite the support of Rabbi Akiva, is interpreted as vindication for Rabbi Yochanan's reasoning. Does history support that conclusion? Perhaps Rabbi Akiva believed that the chosen status of the Jews would overcome both an external threat and an internal dispute. Rabbi Yochanan effectively abandoned the chosen legacy, concluding that an oral tradition, promoted by the rabbis, would best assure Jewish continuity.

This Early Rabbinic Paradigm protected diaspora Jewry at the expense of a covenantal promise. It did so by building ever higher fences, not around the Torah itself, but around those Jews committed to observe Torah law. This necessarily precluded their participation in and benefit from the host society, and limited their impact on the humanities and sciences.[169] The Jewish masses were discouraged from seeking answers on their own or exercising individual, intellectual judgement in even the most mundane aspects of daily life. In his *Changing the Immutable*, Marc Shapiro writes that "R. Hayim of Volozhin (19th century) refused to give an approbation to the *Hayei Adam*, a halakhic handbook by R. Abraham Danzig. He did not believe that the masses should have easy access to texts of Jewish law, and thus no longer be dependent on a rabbi."[170]

---

[169] "Much is made of the opinions of great rabbis of Eastern Europe and Old City Jerusalem of the 1800s regarding the place of some secular studies in the context of Jewish education. I have often wondered what the opinion of those great men would be in twenty-first century society today. Torah and halachic norms are unchanging but Jewish societies and conditions of life have changed considerably over the last three hundred years. Children are entitled to be educated according to the realities of our present world and not according to imagined circumstances of different centuries and locales." Berel Wein, *Jerusalem Post*, 9/4/2009, "The First Day of School"

[170] *Changing the Immutable*, p. 20. Shapiro goes on to say that the Talmudic concept of "ve'ein morin ken," meaning that this is the law but we do not teach it, has often been

Nachmanides (12th century Spain) warned those who make fences around the Torah: they are good, ". . . as long as we all know that this is a fence, and not directly from God in His Torah." As such he disputed the amoraim (the rabbis of the Gemara) who in tractate Yevamot of the Babylonian Talmud (page 21a) said that these fences are in fact part of Torah itself. And this principle served post-destruction society well. Leadership was consolidated in the rabbinate, a self-selected group, well versed in the then rapidly developing Oral Law. In order to standardize understanding and practice the rabbinate chose to violate a scriptural precept by recording the oral tradition. It fixed a methodology of interpretation and went to great lengths to reread history to justify its source, its interpretation and its new standard-bearers.

With Jews spread throughout the world, rabbinically established standards were codified. Innovation, once the cornerstone of Oral Law (whose practitioners were famously respectful of variations in practice among their colleagues), was increasingly limited by a new, overarching and authoritative component of Torah observance: the mesora – a tradition of law and practice inherited through an unbroken generational linkage beginning with the receipt of the Torah by Moses at Mt. Sinai. Since every successive generation of Jews found itself further removed from the original understanding of divine Torah principles, earlier generations became "deified" in the eyes of their successors -- when in doubt, current and future generations must always conform to the practices of their predecessors.

American Rabbi Emanuel Feldman wrote: "In the absence of a Sanhedrin, faithful Jews have always consulted with the"shofet asher yehyeh bayamim ha'hem," the judge who will exist in those future days (Deuteronomy 17:9)." That is, by consulting with universally acknowledged senior rabbinic decisors (gedolim/posekim) whose wisdom, experience, knowledge, and profound Torah learning are indisputable, one can determine what is or is not within the spirit of Jewish law.[171]

The vise of mesora has tightened over time, suffocating the creativity and flexibility with which the oral tradition was created, simultaneously sharpening differences of opinion and practice within the

---

used as a justification for keeping certain knowledge out of the hands of the masses.

[171] *Jewish Thought*, "Preserving our Mesorah in Changing Times," Oct 29, 2010

observant community. Instead of being the source of creative solutions to modern problems or the impetus for the innovative implementation of scientific or philosophical advancement, mesora has become a ball and chain around the ankles of contemporary observance. Consequently, the standing of the post WWII rabbinic paradigm has been diminished, and its proponents have lost the respect of a questioning, well-educated, younger observant constituency. An inclusive paradigm that is willing to look back to scripture as the source of mesora, 1500 years before the rabbis created and began the process of manipulating their own, is gradually replacing it. Rabbi Berel Wein opines, "Professor Chaim Soloveitchik pointed out in a seminal essay decades ago that orthodox Jewry after the Holocaust shifted from a societal based faith group to one that became a book dominated grouping. In this he signaled that the chain of societal tradition that had guided Jewish life for centuries in the Diaspora was now broken, discarded and would shortly no longer even be remembered as having existed at all. And this is a pretty accurate summation of the situation in the Jewish religious world today."[172]

In 2010, Rabbi Hershel Schachter of Yeshiva University raised his shield in defense of rabbinic authority. Those demanding growing roles for women and inclusivity in the conduct of communal worship and leadership had already eviscerated halakhic objections by citing history, Talmudic sources, legal codes and poskim (expert decisors). Rabbi Schachter planted his finger firmly in the growing hole in the dike - articulating a position that allows the law to be superseded by tradition, and upholding the absolute authority of rabbinic Judaism in narrowing the practice and law of Torah observance.

"Mesora is not primarily a corpus of knowledge to master but a process of accessing a chain of student-teacher relationships that reaches back to Sinai. Moses received the Torah and transmitted it to his student, Yehoshua, who in turn taught it to his students and so on, continuing through today." [173] Rabbi Schachter cites Devarim 17:8 as support for the position that when there is uncertainty about law, one is obligated to go to the Central Rabbinic Court, consisting of the greatest Torah scholars

---

[172] Berel Wein, *Jerusalem Post*, Jan 27, 2012

[173] *Jewish Thought*, "Preserving Our Mesorah in Changing Times," Hershel Schachter, October 29, 2010

of the generation. Rabbi Schachter states that a qualified scholar of his generation is the equivalent of one's own primary teacher, the one whose guidance one is obligated to accept. Accordingly, true commitment to the mesora means following the view of one's "Rebbe," or in this case, that of the local scholar/teacher. He cites his teacher, Rabbi Dov Baer Soloveitchik who asserted that Judaism allows for innovation but not change. Innovation must come with the proper motivation, and it is the scholar/teacher and like-minded colleagues who determine the rectitude of that motivation. "Verifying that a new practice is truly in the spirit of Torah requires great breadth of knowledge and depth of understanding of halakha."[174] "This is why many posekim require customs to be those of vatikin, God-fearing scholars whose every action is done for the sake of Heaven." An otherwise proper custom instituted by someone deemed to be unqualified may be ignored. Similar reasoning results in a belief that practices drawn from those of non-Jews are also violations of the mesora. And, changes in synagogue practice are compared to changes in practice at the ancient, holy Temple (citing the Gemara's interpretation of Chronicles 28:19) and would then contravene mesora.

A Modern Challenge to the Mesora

Deuteronomy's command cited by Rabbi Schachter mentions nothing about a Central Rabbinical Court as the interlocutor in a dispute. In fact, when there are matters of dispute "in your cities," you should come, not to a court, but first to the Priests, Levites and judge of those days. And Deuteronomy 31:9 is very specific about the transmission of a mesora: "Moses wrote this Torah and gave it to the Priests, the sons of Levi, the bearers of the Ark of the covenant of Hashem and to all the elders of Israel." When passing along the laws regarding oaths (Numbers 30:2), Moses did not assemble the great Torah scholars of the day, but assembled the heads of the tribes, some of questionable character based on the Torah narrative itself.

The idea that the mesora relies on the structure, composition and role of the ancient Sanhedrin as both model and proof text for the

---
[174] Hershel Schachter, October 29, 2010

leadership template that came after is thrown into doubt by the origin of the very institution that today's rabbinic courts hold dear.

In Numbers 11:16, God commands Moses to gather 70 men who are "elders of the nation, and their officers." Elders occupy a trusted place in Torah as life experiences seem to best mold a leader. In Genesis 50:7, Joseph returns to Israel to bury his father Jacob and is accompanied by not only the elders of his household, but those of the land of Egypt. In Exodus 3:16, Moses is instructed to first tell the elders of Israel about the call to lead the Jews out of Egypt. Proverbs 31:23 highlights the lofty position of the elder: "Her husband is respected at the city gate, where he takes his seat among the elders of the land." Elders were involved in community leadership (1 Samuel 11:3, 16:4, 30:26) and were considered "the wise" (Ezekiel 7:26, Jeremiah 18:18). In Exodus 18, such men are God fearing, trustworthy and honest. The first chapter of Deuteronomy attributes to them wisdom, understanding and knowledge.

The Mishna alters the definition of an elder, and in doing so accelerates the development of early rabbinic Judaism. The new requirements eliminate both life experience and heritage as qualifications. Instead, elders are scholars: "The Zakein is none other than a sage, and the word means zeh she kanah hokhma (one who has acquired wisdom, i.e., wisdom of the Oral Law)."[175]

Sanhedrin is a Hebrew-Aramaic term that referred only to the "supreme court" seated in Jerusalem. It was derived from the Greek word συνέδριον which Josephus used in connection with the decree of the Roman governor of Syria, Gabinius (57 B.C.E.). He abolished the constitution and the then existing form of government of Palestine and divided the country into five provinces. He placed a sanhedrin at the head of each, one of which was Jerusalem.[176] Some think that the term was in use a century or two earlier as it is used in the Greek translation of Proverbs. A council of elders, or a senate, called the gerousia, which existed under Persian and Syrian rule (fourth through second centuries B.C.E.), is considered by some scholars to have been the forerunner of the Great Sanhedrin in Second Temple Jerusalem. The Great Sanhedrin ceased to exist after the Roman conquest in 70 C.E. The

---

[175] Kiddushin, 32b

[176] *Antiquities*, XIV 5 para 4

council later convened in Yavneh where it comprised leading scholars, and functioned as the supreme religious, legislative, and educational body of the Palestinian Jews. Its leader, the nasi, was recognized by the Romans as the political leader of the Jews (patriarch, or ethnarch). This body ceased to function after 425 C.E.

Earlier reference to the Sanhedrin is found during the reign of the Judean king Yehoshafat but does not specify learning as a requirement for membership: "Moreover in Jerusalem, Yehoshaphat appointed Levites and priests, and of the heads of the fathers' houses of Israel, for the judgment of the Lord, and for controversies." (2 Chronicles 19:8) And, according to the Babylonian Talmud (Moed Katon, 26a), King Saul was president of the Sanhedrin during his reign, and his son Jonathan was its vice-president.

Talmudic sources refer to the "Sanhedrin Gedolah hayoshevet be-lishkat ha-gazit" ("the Great Sanhedrin which sits in the hall of hewn stone").[177] Some historians posit that the mention of "Sanhedrin" without the modifier "gedolah" (Jerusalem Talmud, Sanhedrin I. 19c) assumes that another body may have also met in the hall of hewn stone. This may explain why neither Josephus nor the Gospels refer to any decisions of the Sanhedrin that are concerned with priests, the Temple service, or religious law. Instead they discuss only legal procedure, verdicts, and decrees of a political nature. By contrast, Talmudic references to the Sanhedrin in the hall of hewn stone deal with questions relating to the Temple, the priesthood, the sacrifices, and other religious matters. There may have been two high courts operating simultaneously, one that was the supreme political authority and another that dealt strictly with religious law. The religious Sanhedrin seems to have comprised members of the most influential families of the nobility and priesthood.[178]

The pre-exilic books of the Bible make no reference to a Sanhedrin, and as a result the institution was probably not introduced before the time of the Second Temple. It may originally have met only on special occasions for the purpose of considering important questions or issuing regulations concerning religious life. The first assembly of this kind,

---

[177] Sifra, Vayikra, ed. Weiss, 19a

[178] In the Jerusalem Talmud, Sanhedrin IV. 2, there is an allusion to the composition of this body

held under the jurisdiction of Ezra and Nehemiah,[179] is traditionally called "the Great Synagogue" ("Anshei Keneset ha-Gedolah").

The exact composition of the Sanhedrin during the second Temple period is unclear. Early on it reflects the Torah narrative, limiting membership to priests belonging to prominent families under the presidency of the high priest. Later, the Pharisees held prominent positions in this body corresponding to their success in their political and religious conflict with the Sadducees. Near the end of his reign, John Hyrcanus (mid-second century B.C.E.) rejected the Pharisees[180] and eliminated them from the body. A Sadducean court (TB Sanhedrin 52b), or a Sadducean Sanhedrin (Megillat Ta'anit) was then formed. Later, under Alexander Jannæus, Shimon ben Sheṭaḥ ousted the Sadducean members, returning it to Pharisaic domination. They later lost this advantage in a disagreement with Alexander only to regain it under his successor, Salome Alexandra. After the destruction of the Second Temple the Sanhedrin was limited to only sages. The priests were scattered and discredited, the monarchy had fallen, and Torah "sages" had assumed the role of the elders (zekainim).[181] A mesora including the period of the Sanhedrin could not be held up as a model for the transmission of Jewish tradition as its composition and role changed with the political divisiveness and decisions of the day.

According to R. Jose b. Ḥalafta (late first century, early second C.E.), members of the Sanhedrin immediately after the destruction of the second temple were required to be men of scholarship, modesty, and popularity.[182] The Sifre, Numbers 92,[183] indicated that these men also had to be strong and courageous. Qualifications during the Mishnaic period include a candidate having served in lesser offices as a local judge, and later as a member of two lesser courts in Jerusalem. Rabbi Yochanan, a Palestinian amora (a sage included in the Gemara) of the third century listed different qualifications including physical character-

---

[179] Nechemia, VIII-X

[180] *Antiquities*, xvi. 11, § 1

[181] *Wanderings: Chaim Potok's History of the Jews*, p. 191.

[182] Tosefta., Ḥagiga ii. 9; Sanhedrin 88b

[183] Friedman edition p. 25b

istics (tall and intimidating in appearance), advanced age, great learning, conversational ability in foreign languages, and familiarity with the dark arts (Sanhedrin 19a).[184] Qualifications not only varied in the minds of the sages, but more often than not did not emphasize Torah learning.

Maimonides questioned the claim of an unbroken mesora dating back to Moses at Sinai. He believed that the mesora ended with the closing of the Babylonian Talmud (about 600 C.E.). "Our nation is wise and perfect, as has been declared by the Most High, through Moses, who made us perfect: 'Surely this great nation is a wise and understanding people.' (Deuteronomy 4:6) But when wicked barbarians have deprived us of our possessions, put an end to our science and literature, and killed our wise men, we have become ignorant . . . we are mixed up with other nations; we have learnt their opinions and followed their ways and acts. . . Having been brought up among persons untrained in philosophy, we are inclined to consider these philosophical opinions as foreign to our religion, just as uneducated persons find them foreign to their own notions. But, in fact, it is not so."[185] "The natural effect of this practice was that our nation lost the knowledge of those important disciplines. Nothing but a few remarks and allusions are to be found in the Talmud and the Midrashim, like a few kernels enveloped in such a quantity of husk, the reader is generally occupied with the husk, and forgets that it encloses a kernel."[186] "Today many hardships have oppressed us and times have pressured us all, so that the wisdom of our scholars has been lost, and the understanding of those with insight has been hidden, thus those explanations and responses and law that the Geonim (sages of the immediate post Talmudic period) compiled, and considered clear, are now considered difficult to decipher, and only a very few really understand them. How much more so with the Talmud itself . . ."[187]

Is there, then, a true line of the mesora? The Torah calls for Moses to pass its wisdom to the priests and then the elders (zekainim). There is no mention of rabbis or other teachers who might be chosen from among the people as being qualified to become a link in the chain. God ordered

---

[184] *Jewish Enclyclopedia*, 1906 Edition

[185] *(Guide for the Perplexed) Moreh Nevuchim* 2:11

[186] Ibid, 1:71

[187] Introduction to Maimonides' Perush HaMishnah

that His Torah be passed first to those divinely designated for the role of teacher, and then on to those who have the benefit of extensive life experience. Life's lessons and the practical interpretation of Torah took precedence over one's academic training. Even if Rambam is correct (in Hilchot Sanhedrin 4:10) that the beit din (court) of Moses initiated the transfer of semicha (ordination), that transmission is anything but continuous or homogeneous – especially in the opinion of Rambam himself. In fact, many claim that the chain of semicha was broken around 360 CE when Hillel II dissolved the Sanhedrin and fixed the Jewish calendar.[188] If semicha is a mark of masoretic transmission, it was at best terminated 17 centuries ago (Joseph Caro, the author of the authoritative code, Shulchan Aruch wrote that there was no semicha in his day – the 16th century). Maimonides believed that the process and authority of semicha could be renewed in Israel in anticipation of the messianic age. Several attempts to do just that were failures.

The sages speak very little about the existence of God, and their single-minded focus on collegial authority raises as many questions as it answers. Fifteen hundred years after receiving the Torah at Sinai, new leadership of a geographically and ideologically dispersed Jewish nation reinvented the mesora in its own image. "Moses received the Torah from Sinai and passed it on to Joshua; and Joshua passed it on to the elders, and the elders to the prophets and the prophets passed it to the men of the great assembly. They said three things: Be deliberate in judgment, create many students and make a fence for the Torah."[189] The historic, biblical transmission of the Torah was altered. The rabbis substituted themselves for both the priests and the tribe of Levi, and ultimately for the elders.

God instructed Moses (Shemot 34:27), "Transcribe for yourself these matters, for by the mouth of these matters I have signed with you a covenant and with Israel." Here the Gemara explains how the Torah differentiates the written from the oral law. Because each is inherently indispensable in its given form, the written Torah cannot be converted into an oral teaching, and the oral Torah must not be written down. For

---

[188] According to Rabbi Avraham bar Hiyya ha Nassi, Sefer ha Ibbur, 3:7 in the name of Hai Gaon; Also Ramban in Sefer ha Zekhut on Gittin and others

[189] Tractate Avot, Mishna 1, perek 1

fear that the heretofore unwritten knowledge be forever lost to the people, the Gemara (Gittin, 60a) cites psalm 119:126 (Eit la'asot la-Hashem heifeiru toratekha: there is an emergency leaving no other choice) as the justification for the cancellation of this prohibition. Since the recording of the Mishna, written and oral texts are converted to fixed formats that facilitate study. The propagation of this exercise has diminished the exceptional nature of the Law, and by extension, that of both God and the people.

"As the law took hold, by its nature, it crystallized the society. Free expressions died, smothered under a mantle of hyperorthodoxy. Since free thought invited accusations of violations of the law or claims of heresy, prudence, a closed mind, and a silent voice prevailed." This thought actually refers to circumstances 2000 years ago! And yet, the description could not be more contemporary. "Free thought was limited to religious or apocryphal writings, which upheld the orthodox positions of the day. . . Jubilees, Enoch and other apocryphal books found in the Qumran caves are a triumph over the unimaginative mindset brought on by making religious law supreme and human expression contrary to law punishable by death. It may be an odd manifestation that such a burst of creativity was fueled by the very search for order that suppressed free thought in the first place."[190]

The greatest liability imposed by the rabbinic legacy is its effect on the individual's willingness and ability to explore and bolster his personal faith. One thousand years ago, Bachya ibn Paquda wrote of the trap into which the rabbinate has drawn us: Learning about and understanding the unity of God is our responsibility, but for most of us it is "attained by tradition only; the believer having faith in those from whom he has received the tradition not however knowing the truth by the exercise of his own intellect and understanding. Such a person is like a blind man . . . this suggests comparison with a company of blind men, each of whom has his hand on the shoulder of the one in front of him; and so on . . . until reaching a person of sight (but) should this seeing person fail in his duty . . . the misfortune would affect them all. . . So too, if a man accepts the doctrine of the Unity on the ground of tradition

---

[190] *The Book of Jubiliees*, p. 8-9

only he can never be sure that he will not come to associate the worship of one God with the worship of another being."[191]

Our Religions, a book published in 1993, describes the world's seven major religions, each in an essay authored by an expert follower of that religion. Six of these writers couched their descriptive and historical remarks in language of reverence and respect for beliefs that they explained were either divine, traditional or both. The essay on Judaism is presented in language that, by comparison, diminishes the divine origin of the source scripture and its narrative. Jewish exceptionalism and its source are not seen as divine through the lens of academia, nor are they necessarily seen as essential to the observance of the religion itself.

". . . the historical events of 586 and 450 (BCE) are transformed in the Pentateuch's picture of the history and destiny of Israel into that generative myth of exile and return that characterizes every Judaism, then to now."[192] He goes on to say, ". . . systematically speaking, Israel – the Israel of the Torah and historical prophetic books of the sixth and fifth centuries – selected itself."[193] When the Jews returned to Israel after the Babylonian conquest, Jewish history "by definition . . . is invented not described . . . everyone understands the mythopoeic (myth-creating) power of belief . . ."[194]

The transformation of the early rabbinic paradigm from one of rescue to one of self-preservation is manifest in the body of the Alenu prayer. Repeated at the end of each of the three daily Jewish prayer services, this text is ancient in origin. Some believe it was composed by Joshua. Talmudic references (Jerusalem Talmud, Avodah Zarah 1:2) make it likely that Aleinu was included in the Admidah (the silent prayer said standing, which comprises the heart of each daily and holiday service) of Rosh Hashana in the time of the early third century Talmudic sage, Rav. The second paragraph of this prayer includes a reference to "tikun olam," contemporarily understood as "fixing" the world with reference to social justice. It appears, however, that the original text most likely included the variant spelling "tachen" rather than the current

---

[191] *Duties of the Heart*, Volume 1, p. 63-65

[192] *Our Religions*, p. 310

[193] Ibid, p. 315

[194] Ibid.

"takein." In the former, the intent is that God's sovereignty will be established over the world -- emphasizing if you will, the exceptional role to be played by the Jews in doing so. Contrast this with mankind's current charge to husband social justice in its repair of the world or one's individual character as directed by the rishonim (rabbis of the early Medieval period) and kabbalistic texts.

Meir Bar-Ilan[195] suggested that the spelling is that of the original text (l'tachen rather than l'takein). The l'tachen text is found in the siddur of Sadiah Gaon (tenth century) and later in the text of the Rosh Hashana Amidah in the Mishneh Torah (the legal code of Maimonides in the 12th century). This is also the text to this day in Yemenite siddurim (prayer books). The Ashkenazi reading since the time of the rishonim has been l'takein (going back to at least the Mahzor Vitry of Rabbi Simhah of Vitry, 12th century) and Siddur Hasidei Ashkenaz of the same period. Michael First explains that the original reading was likely l'tachen because of the content of the second paragraph of the prayer: "Beginning with the second line. . . every clause expresses a hope for either the removal of other gods or the universal acceptance of our God. With regard to the first line, properly understood and its mystical and elevated language decoded, it is almost certainly a request for the speedy rebuilding of the Temple. Taken together, this whole section is a prayer for the rebuilding of the Temple and the establishment of God's kingdom on earth. This fits the reading l'tachen perfectly."[196]

By the 12th century, the rabbis had largely given up on the universalist, exceptionalist biblical message and had settled while in exile on the individual, corrective focus. This conclusion is supported by First's belief that the two sections of the Aleinu prayer were composed simultaneously. He reasons that the earliest texts are not split into separate paragraphs, both sections quote or paraphrase the same chapter of the prophet Isaiah (chapter 45), and terms characteristic of heikhalot (mystical) literature are found in both sections.

---

[195] Mekorah shel Tefillat 'Aleinu le-Shabeah,' Daat Vol.43, Ramat Gan, 1999, p. 20, n. 72 (cited in "Aleinu: Obligation to Fix the World or the Text?" Mitchell First, Hakirah, p. 187

[196] Ibid, p. 195

It may be that as rabbinic Judaism developed, powerless under the privations visited on them by their hosts, and divorced from the exceptionalist idea, the expansion of the notion of l'tachen haolam became l'takein olam. The collective "establishment" of the sovereignty of the God of the Torah was replaced as a Jewish responsibility by the piecemeal correction (as derived from heikhalot literature) of an individual through introspection. Correcting one's individual shortcomings was a task more easily undertaken than insisting on having the Jewish people lead a global renaissance in thought and action.

Arnold Jacob Wolf writes, "All this begins, I believe, with distorting tikkun olam. A teaching about compromise, sharpening, trimming and humanizing rabbinic law, a mystical doctrine about putting God's world back together again, this strange and half-understood notion becomes a huge umbrella under which our petty moral concerns and political panaceas can come in out of the rain,"[197] It should be noted that Joseph Karo is a key figure in the Kabbalistic school of thought which developed the concept of Tikkun Olam. He is also the author of the Shulchan Aruch, the purpose of which is to facilitate individual observance of halakha (Jewish law) which is the way to repair the world.

It seems that the earlier language of tachen olam is aimed at bringing all Jews together in order that they lead the world toward an understanding of God and His sovereignty. The pursuit of correction, of righting social wrongs, has never been pursued by more than a narrow portion of the community. And on each occasion even the most salutary results are short lived. Many Jews were instrumental in the success of the American civil rights movement, yet fifty years later Black activists target Jews as their oppressors. Jews led the feminist movement but today are excluded from speaking at their rallies. A national march for racial justice is deliberately scheduled for Yom Kippur, excluding most Jews from attending.

By paying little more than lip service to the chosen designation, the rabbinate has discarded this aspect of Jewish leadership. The original meaning of the Aleinu prayer seems irretrievable. The mesora has been assembled in the absence of one of the founding principles of the Torah and has become unquestionably accurate and unchallengeable.

---

[197] *Judaism* 50:4, Arnold Jacob Wolf, "Repairing Tikkun Olam"

"In the hierarchy of Jewish law, custom ranks at the bottom. It carries neither the force of biblical nor even rabbinic law, and when one considers the horrific consequences of religious fighting, to create dissension over such is almost idiotic."[198] And the parsing of mesora goes beyond custom or behavior. According to Samson Rafael Hirsch, "Torah originated with what God told Moses; Aggada originated in the mind of a sage."[199] Rabbi Moshe Shmuel Glasner, the great grandson of the great Hungarian sage, the Chatam Sofer, objected to the "fixed" oral law. In his *Dor Revi'i* he wrote that the written Torah is permanent, but the oral Torah is not, nor was it intended to be. God wanted each generation to apply Torah to its own circumstances, so it had to be passed orally from generation to generation. Were that the case today, precedent would play only a small role. But writing down the oral law (the Mishna) had the effect of narrowing and limiting its application.

"Somehow we have rewritten the past in order to make it fit our current normalcy," says Berel Wein. "Doing so is dishonest and dangerous for it creates a world that never existed and gives false answers to the problems of previous generations. It excuses their failures and prevents a clear assessment of our current challenges."[200]

The rabbinate conducts a campaign not altogether different from that of a political party. Columnist Joe Klein of *Time Magazine* describes the unusual political philosophy of an American presidential candidate in "repurposing the past as the future. In the end, though nostalgia is a sepia-toned refuge for those suffering a sense of diminished capacity . . . It is a nursing home for those more comfortable looking back than looking forward."[201]

The rejection of the "chosen" aspect of Jewish heritage – the fundamental, original and divinely dictated, defining characteristic of the Jewish nation -- has discouraged the assertion of the unique leadership role intended for a unique people. The ascension of mesora is both an expression of the rabbinic repurposing of Jewish exceptionalism and the

---

[198] Torah in Motion, Kelman, 8/18/13, Pesachim 50

[199] HaMaayan, Letter on Aggadah, 1976

[200] Rabbiwein.com/blog/post-1889.html?, 5/25/16

[201] *Time Magazine*, May 30, 2015, p. 29, "Trump, the astute salesman, has seized on America's prevaliling mood: nostalgic"

mechanizing of faith. S.T. Coleridge noted almost two centuries ago that from the point of view of the other nations, "The religion of the Jew is, indeed, a light; but it is as the light of the glow-worm, which gives no heat, and illumines nothing but itself."[202] Berel Wein goes even further: "Not believing in the eternal truth of our own God given cause has led us to lose our standing in the eyes of the world."[203]

Biblical leadership required both empathy and vision. Empathy was acquired and exercised in the acquisition of life experiences and was embodied by the zakain (elder). Unapplied learning served no purpose in scripture. There were no "ivory tower" graduates among its elders. Yet the rabbis of the Mishna were quick to replace those elders with men dedicated exclusively to Torah learning.

Spiritual leadership is provided by those who advance the well-being of the community while simultaneously inspiring those within and earning the respect of those without.

---

[202] *Table-Talk*, S.T. Coleridge, April 13, 1830

[203] *Jerusalem Post*, Berel Wein, August 7, 2009, "No Free Lunch"

VI

THE BIBLICAL CHOSEN PROMISE

DEFINING EXCEPTIONALISM

The Hebrew Bible describes monotheism as a divinely offered improvement over the polytheistic practices of the ancient world. This new religious ideology was provided by a single, all-powerful God who declared Himself the creator of all existence. Scripture was given by that one God to a particular people -- a community of families -- who having been chosen, accepted and carried out the obligations dictated by that text.

According to the chronology of the written Torah/Pentateuch, its directives and laws including the chosen/exceptional assignment, were presented to and accepted by that community 1312 years prior to the Common Era.

In presenting the biblical text its author identified five components exclusive to the "chosen" assignment:

- **The divine scripture given only to the 12 tribes of Israel,**
- **The unique and fertile Hebrew language in which that scripture was transmitted,**
- **The natural science of creation and the underlying Kabbalistic tradition that explains it,**
- **The Land covenantally promised to only this nation, and**
- **The messianic idea.**

The Five Books of Moses lay out the story of mankind's confrontation with a supernaturally created reality. But what exactly is that reality? Is it an elaborate fiction – not reality at all? Is it a living history as we have come to know it, or is it just a part of some greater, hidden and more complex reality? Man seems to be the protagonist in a timeless mystery. We have, at first slowly, and more recently very rapidly come into the possession of tools that allow us to unravel that mystery, much to the mutual chagrin of scientists, philosophers and clerics.

One of those tools is archaeology. Another is physics. And within the latter category is found the concept of entropy. We will soon see how its application to religious tradition and thought can provide an understanding of religious topics from free will to ritual impurity. Knowing something about entropy, however superficially, will allow us a glimpse of what God had in mind when He chose the Jews.

> *Sidebar:*
>
> *The concept of entropy is fundamental to thermodynamics -- the branch of physics that describes heat, temperature, energy and work. Entropy is commonly associated with the idea that everything in the universe routinely moves from a state of order to one of disorder. Entropy is the quantitative measurement of that change.*
>
> *Ice melting in water illustrates entropy in action. There is a change from something with a fixed form to something "unrestricted" so that the disorder of the scene being observed has grown – its entropy has increased. But entropy is also a characteristic of something very "immaterial" – it is a principle observed in information theory that is related to probability.*
>
> *The dictionary provides a concise statement of the above two definitions of entropy: 1) a thermodynamic quantity representing the amount of energy in a system that is no longer available for doing mechanical work,*

and 2) a numerical measure of the uncertainty of an outcome in communications theory. Both of these definitions describe the glue that holds history, language, biology and the science of creation together. So, stay tuned.

The remainder of this text is an argument for the only body of practice, text and ideology —Judaism —that blends scripture and science. The text will speak to the co-opting of the Jewish narrative by later monotheistic faiths. This process has diminished the primacy and understanding of the historical record, the "miraculous" power of the Hebrew language, and the monism of science and mysticism. Reacquaintance with these components of the exceptional role can release Judaism from the catholic limitations copied and adopted from its successor religions.

Is there a God? Where do we come from? Why are we here? Where are we going? Why do bad things happen to good people? Do we have free will? If God is good, why is there evil in the world?

Mankind's institutional religions preach the impossibility of divining answers to these fundamental questions. Faith should suffice. Do we have permission to know the "mind of God?" Is the pursuit of this knowledge in some way an affront to the Creator? On the contrary, scripture is challenging us to answer these questions, if not for mankind as a whole, then for each of us individually. Advancements in science, philosophy and archaeology demand the application of innovative thought to our past, our present and our future.

That analysis begins with scripture.

SCIENTIFIC AND RELIGIOUS PARADIGMS

The terminology that we routinely use to describe watershed, historic change is drawn from science. Yet, science and religion are so interdependent that the fundamental concepts of each are almost indistinguishable from one another. How has science analyzed its historic development and change, and how can the same analysis be applied to religion?

In 1962, Thomas Kuhn coined the word "paradigm"[204] to describe "universally recognized scientific achievements that for a time provide model problems and solutions to a community of practitioners." Scientists were accustomed to applying models that dictated just how they would view the world around them. So, periodically, when anomalies arose concerning these models, the resulting "revolution" triggered the establishment of new paradigms, which then dictated a new and different type of procedural thought.

Kuhn understood that paradigms dominate "normal science." Normal science explores solutions within an accepted framework; the rules of the game have already been established. Yet, new paradigms are routinely introduced and ultimately accepted.

For example, in the early twentieth century Newtonian physics was modified by Einstein's relativistic worldview. In 1930 the existence of the neutrino was postulated. Its existence was taken for granted for 26 years before it was actually discovered experimentally. On the other hand, the geologic idea of plate tectonics was not accepted by the scientific community for almost half a century. Only the introduction of novel experimental technology convinced researchers of its validity.

Paradigm shifts take time.

In the years before the Second World War, the scientific community accepted a new paradigm called the Verification Principle (that statements are meaningful only if verified by sense data). Shortly after the war that paradigm was replaced by Empiricism (science starts from publicly observable data, independent of theoretical assumptions). Later, advances in quantum mechanics replaced the understandings of classical physics at sub-atomic levels. Stephen Worfram explained in a 1984 *Scientific American Magazine* article that "physical systems are viewed as computational systems, processing information much the way computers do. A new paradigm has been born."

Many see the same spirit of inquiry, analysis of change, and template for thought in religion that they see in science.[205] For example,

---

[204] *The Structure of Scientific Revolutions*

[205] *Myths, Models and Paradigms*

Hans Kung[206] is one of many who have attempted to explain the development of Christian theology in terms of a series of major paradigms, each representative of an historic period.

It is in this spirit that "Religious paradigms, like scientific ones, are not falsified by discordant data but replaced by promising alternatives."[207] Religious paradigms are generally understood to be traditions established and historically perpetuated. Identical criteria are employed by historians to assess both scientific and religious theories, and they are very much a product of common sense: simplicity, coherence, and agreement with experimental evidence. A scientific theory is credible if it accurately accounts for what has been observed and gives us accurate predictions of future measurements. But common sense has always been heavily influenced by one's point-of-view. Data is not as pure as we would like to think that it is. It is always burdened by theory, strongly influencing one's choice.

Kuhn (in *The Structure of Scientific Revolutions*) made this point a cornerstone of his paradigm development. He recognized that observational data and criteria for assessing theories are paradigm-dependent. As a result, scientific choice is not objective at all -- it is entirely subjective and dependent on the methods and beliefs of a particular scientific community. This is very much a "religious" conclusion.

Ian Barbour, professor of physics, professor of religion and Gifford Lecturer wrote, "A shared paradigm creates a scientific community – a professional grouping with common assumptions, interests, journals and channels of communication. This stress on the importance of the community suggests parallels in the role of the religious community . . ."[208] Barbour explains that for the community to function, scientific anomalies are set aside or accommodated by modifications to stay within the accepted paradigm. But when there are too many, the scientific community examines its assumptions and searches for alternatives. When that happens, a new paradigm causes a revolution in thought and behavior that is incompatible with the old. As we see anomalies building in religious practice and thought, a similar revolution follows.

---

[206] *Christianity*

[207] *Myths, Models and Paradigms,* Chapter 1: Introduction.

[208] Ibid, Chapter 6

Judaism is undergoing a paradigm shift, just as it has on several previous occasions. "Anomalies" are increasingly not being "set aside or accommodated by modifications to stay within the accepted paradigm." As will be seen below, there are now "too many" anomalies and the religious community must "examine(s) its assumptions and search for alternatives." The paradigm shift is being vigorously resisted by the legacy holders of the previous paradigm, yet the foundation of that paradigm has become so weak that action to facilitate and implement change is an accepted part of the fabric of 21st century religious Jewish life.

"As scientific models lead to theories by which observations are ordered, so too do religious models lead to beliefs by which experiences are ordered. Beliefs, like theories, can be propositionally stated and systematically articulated. But can religious beliefs be tested against human experience, as scientific theories can be tested against observations? Are there any criteria for the assessment of religious beliefs?"[209] In fact, religious beliefs are tested against human experience all the time. The new paradigm on which Judaism will settle will be the result of decades of testing, fine-tuning and compromise.

HISTORY'S FIVE JEWISH PARADIGMS

Jewish history began more than 3300 years ago with the creation of a singular people or nation. It saw the dispersal of that nation to a broad diaspora, leading to the imposition of social and legal constructs designed to preserve a unique national character while in "exile." Ultimately it found that nation reassembled in its scripturally designated homeland. That history comprises reactive paradigm shifts, each developed and sustained primarily for survival purposes. The paradigm shift now underway is unique in its spiritual and developmental motives rather than for its focus on survival.

The Priestly Paradigm describes how the Jews were first gathered together as a distinct people, later escaped Egyptian bondage, found their way to the Promised Land, conquered that land, created a commonwealth there in which religious and civil authority were clearly divided, survived national destruction and exile, and witnessed a return

---
[209] Ibid, Chapter 7

within a century to reestablish that commonwealth. The Jews were designated as a special, "chosen" people. Their exceptional status was to be permanent and the impetus for a new world order that would accept universal recognition of that status. The scriptural and legal stratification of this "exceptional" society left the priests with enormous power over both the spiritual and social life of the nation as a whole, while the king exercised civil control.

The *Priestly Paradigm* was shaken during the Second Commonwealth by the Hasmoneans, the ruling family that united the priesthood and monarchy. That paradigm imploded at the destruction of the Second Temple when much of the priestly tribe became unredeemingly corrupt while others were forced to flee the city of Jerusalem and gather in outlying settlements like Qumran. The newly created variants of the existing paradigm were too many and too disruptive to leave the status quo in place. The Hasmoneans ignited a process that led to the disintegration of both the priestly tribe and its leadership.

From the violent destruction of the Priestly Paradigm emerged the *Early Rabbinic Paradigm*. The rabbinic cult sought to fill both the priestly and political voids by establishing themselves geographically (Yavneh) and authoritatively. Their leadership narrowed the exceptional role granted the Jews in scripture by debating, fixing and recording the oral law, establishing it as the authoritative guide to daily Jewish life. The Tanaaim (rabbis of the Mishna) and their successors, the Amoraim (the rabbis of the Gemara), Savoraim and Gaonim (post Talmudic sages) established a legal and philosophical methodology designed to connect and preserve the diaspora communities. Building on this foundation, Rishonim (early medieval commentators) organized the oral law into written codes, further standardizing both individual and communal conduct and simultaneously reinterpreting Jewish exceptionalism as an internal characteristic of a dispersed, homeless nation.

The Early Rabbinic Paradigm began to crumble under the weight of religious, social and political oppression by both the Christian and Muslim rulers in whose lands Jews found themselves. Repression and hopelessness gave rise to false messiahs who briefly inspired and energized the rank-and-file. Neo-Kabbalistic thought became widespread, and the undereducated looked to local spiritual leaders to provide hope for both their personal and communal futures. Chassidic movements proliferated, building on the words of 16th century Kabbalists and compensating for widespread illiteracy.

After the unsuccessful "messianic" movements led by followers of Shabbetai Zvi, Jacob Frank, and others, the Early Rabbinic Paradigm was in tatters. The Baal Shem Tov and other Eastern European Chassidic masters took whole communities in hand, substituting their personal, divine connection for that of their individual followers. This transitional period and its splintered leadership preceded the Enlightenment, a period during which Jews were permitted to participate in the commercial and intellectual lives of their host diaspora societies. But this new found freedom to integrate became an existential threat to the rabbinic model. For the first time, large scale assimilation became a consequential danger to Jewish survival. The rabbinic cult struggled to offer alternatives that might counteract the effects of these new freedoms, settling instead on the aggressive rejection of ambient society as a primary defense mechanism. Tightened behavioral norms were manifest in the form of law. Contact with new ideas and heretofore unavailable lifestyles was minimized or prohibited in order to preserve tradition and insure the survival of the increasingly narrowly defined oral law. In this *Anti-Enlightenment Paradigm*, the religious core was to be preserved while broader intellectual exposure and understanding were sacrificed.

This paradigm dominated religious Jewish life for almost two centuries, succumbing to the near complete annihilation of European Jewry in World War II.

A miraculous re-emergence of traditional, observant Judaism was spawned by the Holocaust. While generations of Jewish religious life, leadership and norms had been erased in just a few short years, a new paradigm -- that of *Reimagined Judaism*-- became the model that persists until this day. Observant Judaism took on the character of what its young practitioners imagined pre-war tradition and observance to have been. The born-again (baal tshuva) Jews, now "returning" to their religious roots, were really returning to a fiction, created out of whole cloth.[210] Education, the conduct of daily affairs, secular interactions and

---

[210] *Tradition Magazine*, "Rupture and Reconstruction: The Transformation of Contemporary Orthodoxy," Haym Soloveitchik, Vol. 28, No. 4, Summer 1994

even styles of dress were institutionalized and became requirements of "belonging."

The twenty-first century sees alternative forms of religious lifestyle, education and Zionism proliferating. Reimagined Judaism is fraying at the edges. The financial and social health of cloistered, insular communities has become increasingly uncertain. Financial support from the parents and grandparents of young Jews immersed in this environment is shrinking. The ability of these young people to offer the necessary economic support to their offspring is unlikely without social and commercial compromises. Though minimized by much of the rabbinic community, the connection between Zionism and redemption is as clear to many as the existence of the State of Israel. The religious education and changing role of women within observant Judaism has never been at the same time so accepted and yet so contentious. The beginnings of a paradigm shift are apparent in communities where right-wing, left-wing and centrist institutions and practitioners have begun "cross-fertilizing." Institutions are widening their appeal to both retain their traditional bases and attract new adherents. Women have become empowered and their participation in traditional observance is expanding. The *Inclusivity Paradigm* is increasingly at odds with the institutions that doggedly hold themselves apart and have traditionally represented the community to both members and outsiders. As alternatives to the old paradigm consolidate, traditionalists and their rabbinic leaders vigorously defend their positions while finding themselves increasingly disrupted by the progressing paradigm shift.

Can the Inclusivity Paradigm, by virtue of its democratic nature, husband a rekindling of the biblical Jewish exceptionalism discarded with the collapse of the Priestly Paradigm? Will the new paradigm allow traditional Judaism to accommodate pluralistic movements, revitalizing a long-missing and broader sense of Jewish community?

The exceptionalism conferred upon the Jews in scripture defines both God's promise as well as paths to Jewish independence, self-sufficiency and leadership that have as yet been only partly realized.

If God exists and the Torah is divine, then Jewish exceptionalism is as valid as the text that describes it. Jews *must* be the chosen people. If

Jewish exceptionalism can be parsed and each of its parts can be shown to characterize the uniquely Jewish birthright and its inevitability, then the power of that "chosen" status and its exclusivity becomes tangible. In exploring these features, we will see that centuries of Jewish misery and underachievement and an historic willingness to be considered one of three equally legitimate "monotheistic" faiths highlights a nearly universal ignorance of these exceptional, uniquely Jewish divine gifts. Distinguishing Judaism from its sister faiths begins with an understanding of its unique history as recorded in scripture – a history that is woven into and dependent upon the development and implementation of both language and science.

VII

THE ELEMENTS OF JEWISH EXCEPTIONALISM

1. SCRIPTURE, ITS AUTHENTICITY AND HISTORICAL ACCURACY

How can the Five Books of Moses that are in our possession today be the same as those given to (or recorded by) Moses at Sinai 3300 years ago?

It seems hard to imagine that generation after generation succeeded in maintaining the authenticity and accuracy of the text for thousands of years -- yet this is the claim of Jewish traditionalists. Arguments to the contrary include the Documentary Hypothesis (a literary analysis of scripture that attributes its text to several authors) as well as those that center on later editing done by the exilic leader Ezra. Professor David Weiss Halivni, for example supports this latter hypothesis citing the biblical books of Ezra and Nechemia, as well as the Midrash and Gemara as sources for changes and adaptations to the text.

Yet, the Midrash is also the source of this traditional belief: "Before his death, Moses wrote 13 Torah scrolls. Twelve of them were given to each of the tribes. The 13th was put in the Ark (which also contained the tablets from Sinai). If anyone would try to rewrite or change the Torah, the one in the Ark would be evidence against him."[211] It is believed that in Temple times an accurate text was always present. Similarly,

---
[211] Devarim Rabba 9:4

tradition states that an authoritative copy also accompanied the Jews into their Babylonian and Roman exiles. Rabbi Akiva Eiger in his gloss on Tractate Shabbat, page 55b lists 20 places where today's Torah text differs, sometimes by just a single letter from that of the Amoraim (rabbis of the Talmudic period), Rashi (medieval commentator) and the Tosaphists (later medieval commentators).

The Talmud lists almost two dozen features that qualify a Torah as authoritative and, not surprisingly, there is amazing consistency today in the texts of Torah scrolls of all ages and sources. Each has 304,805 letters and only Yemenite scrolls have any differences in content at all-- 9 letter differences in total, none of which affects the meaning of the words. The Samaritans follow a Torah written in their own alphabet, containing as many as 6000 variations from the traditional Jewish Torah text. Samaritans believe that the Jews separated from them in the 11th century B.C.E. Jews however, generally believe the Samaritans to be descendants of the tribes that were relocated to Israel by the Assyrian conquerors of the Northern Kingdom in the 8th century B.C.E. Some historians believe that the Samaritan community was an offshoot of those Jews taken to Babylonia after the destruction of the First Temple. Still others believe that the schism took place around the time of the Hasmoneans. In any case, both texts record the selection of the Jews by God, and in the same language.

Even if the text of the Torah is exactly as it was 3000 years ago, only proof of its divine origin can validate the exceptionalism that it bestowed upon the Jews. Proof in the legally accepted sense can never be discovered as there are no witnesses to call on. Nevertheless, the revelations of natural science, archaeology and linguistics now conspire to present a convincing argument for the belief that the Torah that we read today does in fact have a divine source.

BIBLICAL AUTHORSHIP

Harold Bloom[212], picking up on the 19th century work of German Bible scholar Julius Wellhausen, presents the classic, humanistic ap-

---

[212] Harold Bloom is also the author of the *Book of J,* which supports the documentary analysis of the Torah, identifying it as a book developed over time by many authors and later edited by one or two individuals of decidedly un-divine stock.

proach to biblical authorship in his introduction to *The Western Canon*. A prolific academic and writer, his career has taken him from literary criticism to the humanistic, literary disassembly of the Bible and monotheism in total. His work discusses the products of 26 of the Western World's most important authors. In his preface he describes the "strangeness" of originality that distinguishes the authors that he has chosen. He saves his most enthusiastic comments for the author of the world's oldest work of fiction:

"After Shakespeare, the greatest representative of the given is the first author of the Hebrew Bible, the figure named the Yahwist or J by nineteenth-century biblical scholarship. J, like Homer, a person or persons lost in the dark recesses of time, appears to have lived in or near Jerusalem some three thousand years ago . . . Just who the primary J was, we are never likely to know . . . I am happy to adopt the suggestion belatedly [of a reviewer of his *The Book of J*]: Bathsheba, mother of Solomon, is an admirable candidate. Her dark view of Solomon's catastrophic son and successor, Rehoboam, implied throughout the Yahwistic text is thus highly explicable; so is her very ironic presentation of the Hebrew patriarchs, and her fondness both for some of their wives and for such female outsiders as Hagar and Tamar. Besides, it is a superb, J-like irony that the inaugural author of what eventually became the Torah was not an Israelite at all, but a Hittite woman." [213]

Bloom goes on to say that Bathsheba (J) authored Genesis, Exodus and Numbers. He notes that later censors and revisionists like Ezra didn't like the all too human portrayal of God offered by Bathsheba.

"He eats and drinks, frequently loses his temper, delights in his own mischief, is jealous and vindictive, proclaims his justness while constantly playing favorites, and develops a considerable case of neurotic anxiety when he allows himself to transfer his blessing from an elite to the entire Israelite host. By the time he leads that crazed and suffering rabblement through the Sinai wilderness, he has become so insane and dangerous, to himself and to others that the J writer deserves to be called the most blasphemous of all authors ever."[214]

---

[213] *The Western Canon*, Introduction

[214] Ibid.

Bloom seeks maximum impact with his conclusion: "The ultimate shock implicit in this canon-making originality comes when we realize that the Western worship of God -- by Jews, Christians, and Muslims -- is the worship of a literary character, J's Yahweh, however adulterated by pious revisionists."[215]

Bloom follows a two century-long line of biblical literary critics. Interestingly, though, in the 3000 years of Jewish existence leading up to the "discovery" of biblical criticism ". . . the Torah was investigated from all possible angles by scholars, philologists, masters of the Halakha and masters of the Aggadah, by philosophers and historians, jurists and archaeologists, believers and non-believers, by Jews and Gentiles. All of them found in the Torah material for research, soluble and insoluble problems, but no one found ground for pulling it to pieces. And yet during the past two centuries it has become evident that it is impossible to understand Scripture without the five different sources and at least two editors."[216]

Early "documentarians" chose to interpret the Old Testament as if it had four authors, mirroring the structure of the New Testament. Some of their rationale for multiple human authorship is based on the use of several different names of God within the body of the Bible. Much is also based on varying literary styles from book to book or chapter to chapter -- notwithstanding the fact that the same could be said of Shakespeare or any other author who was known to use a variety of writing styles both within and across works.

Among responses to this literary criticism is the notion that critics ignore personal context, that the words of the Torah are the words of its speakers. "Moses, just before his death at the age of 120 would naturally use different language from the Moses of 80 who brought the people out of Egypt. He employed one style in addressing the people who witnessed the miracles and yet proved rebellious and complaining, and a different style for the new generation, born in the wilderness, and preparing for entry into the Promised Land."[217]

---

[215] Ibid.

[216] *The Modern Jew Faces Eternal Problems*, p. 256

[217] Ibid, p. 258

The Torah is also quite repetitious. Versions of the same material, with notable discrepancies, is often found in multiple sections of the text. And those discrepancies point to a unique emphasis in each version – whether it is a new and different aspect of holiday observance, or the focus of the content of the Ten Commandments. Aron Barth offers, "Is it likely that one and the same author would have repeated the same commandments with differences in language and expression? The answer is that this is possible only in the case of one author. Had there been different sources combined by one editor he would surely not have allowed such discrepancies . . . To accord with the opinion of the critics, the editor would have had to have been a forger of the first order; a man who assembled various sources of as many periods, written without historical responsibility by poets and spinners of legends, sources known exclusively to him. For otherwise his forgery would have been brought to light at once."[218]

Bible critics see the clash of style and subject between the books of Exodus and Leviticus as proof of their different authors. But, for example, how much weight do they give to the fact that the Tabernacle (mishkan) serves as a primary focus for both? The book of Exodus is concerned with the function of the Tabernacle and how it is to be built, while the book of Leviticus has the complementary concern of how it is to be used. The book of Exodus describes the scene at Mt. Sinai at the giving of the Ten Commandments. The book of Leviticus instructs in the use of sacrifices that parallel that event and tie the ongoing use of the Tabernacle to that seminal event in Jewish history.

There are a variety of technical analyses of the books of the Torah that demonstrate structural nuances tying them together. In his commentary on the portion Behar, Rabbi Menachem Liebtag[219] demonstrates the unified structure and themes of Exodus and Leviticus. He shows that 17 discrete units of commandments stretching back into Exodus and forward into Leviticus center logically around the dual themes of the book of Leviticus, the dwelling of the presence of God on the Tabernacle and its effect on the nation.

---

[218] Ibid.

[219] Torah Study Center, Menachem Liebtag, Parshat Behar building on the construct of Rabbi Yoel Bin Nun of Yeshivat Har Etzion.

Other literary patterns appear consistently and are described by both academic and religious researchers, and support the argument for a single author:

"The subject of this composition is the literary model three-four in the Bible, meaning literary units built in four layers. The first three repeat one another and there is not usually a monumental change from verse to verse, and only in the fourth unit begins the severe change, this change which is the central and climactic part of the literary unit."[220]

And while not all of the Biblical critics believe that there were multiple authors and editors, all are satisfied that those authors were just humans. ". . . the first nine books of the Hebrew bible are a unity . . . being the invention of a single great exilic mind, a mixture of national religious curator, seer, historian, priest (however much help he may have had from his confreres, and however much small details of his work may have been tinkered with by later Pecksniffian minds) . . ."[221]

British Bible scholar, historian and Egyptologist Kenneth Kitchen describes the agenda-driven research approach and literary criticism that may be applied here. "To put it a little brutally and inelegantly, one load of wholly imaginary codswallop is worth no more than any other, differing load of wholly imaginary codswallop, and without a satisfactory factual basis none is worth much or perhaps anything at all."[222]

It is interesting to note that the question of authorship of seminal texts is not limited to the Bible. Retired U.S. Supreme Court justice, John Paul Stevens questioned whether William Shakespeare actually authored the many works for which he is assigned credit. Stevens argues, "The historical William Shakespeare was a commoner with no more than a grammar school education . . . where are the books? You can't be a scholar of that depth and not have any books in your home. He never had any correspondence with his contemporaries; he never was shown to be present at any major event – the coronation of James

---

[220] "The Pattern of the Numerical Sequence Three-Four in the Bible," Yair Zakovitch, Makor, 1977, 132-139

[221] *Surpassing Wonder*, p. 24.

[222] *Windows into Old Testament History*, "Assessing the Historical Status of the Israelite United Monarchy," Kenneth A. Kitchen, p.128

or any of that stuff. I think the evidence that he was not the author is beyond a reasonable doubt."[223]

HISTORICITY OF THE BIBLE

"Said to be somewhere along the river Euphrates, the Garden of Dilmun was where Babylonians believed that mankind was created. The similarities between the Dilmun epic and the Garden of Eden story found in the book of Genesis are too similar to be ignored."[224] The giving of the law to King Hammurabi by the sun god Shamash – text on an 8 foot black stone – has 50 of the commandments of the Torah, "practically verbatim."[225] And, the epic Gilgamesh describes an event almost identical to the biblical flood.[226]

If events recorded in the Bible are also common to other cultures, does that diminish the veracity of the biblical narrative or testify to it? Does the claim of antiquity of events by the Biblical Documentarians negate the word of scripture?

Defending the Bible against its critics is in great part a defense of its historical accuracy. After all, if the biblical narrative is fiction, its potential divinity is irrevocably shaken. Rebuttals to those challenges have until recently rested on little more than circular reasoning and common sense. For example, Barry Beitzel, former professor of Old Testament and Semitic Languages at Trinity Evangelical Divinity School has reasoned, "Why . . . should one expect to find the names of an obscure nomad (Abraham) and his descendants in the official archives of the rulers of Mesopotamia? These are 'family stories,' says Beitzel, not geopolitical history of the type one might expect to find preserved in the annals of kings."[227]

---

[223] Scientific American, "Shakespeare, Interrupted," Michael Shermer, August2009, p. 30. Quoting *The Wall Street Journal*, April 18, 2009.

[224] *The Moses Legacy*, p. 5-6

[225] *A History of the Bible*, p. 37

[226] Deluge tablet: Gilgamesh, Sumaria, 2750 BCE. Utrapishtim builds an ark; warned by the gods.

[227] *Is the Bible True*

The best known ancient histories neither mentioned the miracles of Egypt nor the Exodus. But historians recognize that ancient, written histories were typically public relations pieces for the empires of the day. To find no mention of military defeats or divine plagues should be expected.

Nahum Sarna, professor emeritus of Biblical Studies at Brandeis University, reasons that the Exodus story "cannot possibly be fictional. No nation would be likely to invent for itself . . . an inglorious and inconvenient tradition of this nature, unless it had an authentic core."[228] Similarly, Biblical scholar, Richard Elliott Friedman notes that any self-respecting, culture could certainly come up with prouder origins than did the Jews. They would likely portray themselves as "descended from gods or kings, not from slaves."

Additionally, and unlike all other ancient societies, the ancient Israelites had no inherited hierarchy. The Exodus was an "equalizing" event in the words of Rabbi Dr. Joshua Berman. Where in other societies gods spoke only to kings or prophets, at Sinai God spoke to the masses. And the Torah directs the people to perpetuate this history. "It is in the Torah that we see for the first time the realization that the identity of a people may be formed around an awareness of its past. Indeed, the Hebrew Bible is the first literature before the Hellenistic period that may be termed a national history . . . Although there are over one million inscriptions in our possession from the ancient Near East, there is nowhere evidence of a national narrative that a people tells itself about its collective, national life, of moments of achievement or of despair, recorded for posterity."[229]

The national stories of the ancient Near East testify to battles of the gods and kings, and were typically discovered in temples. These accounts were recorded for later royalty, not for the common man. Cambridge anthropologist Jack Goody noted that a culture's willingness to share its religious literature reflects an emphasis on the individual within that culture. And that is characteristic of the Pentateuch where the power rests in the hands of the people -- they observe and administer the law, they appoint a king, they appoint judges. The Torah also divides

---

[228] Ibid.

[229] www.Aish.com, Joshua Berman, "How Torah Revolutionized Political Thought"

power between the king, the judiciary, and the priests. And taxes (tithes) were ordered to be distributed to the needy.

Unlike other ancient histories, biblical descriptions of the lifestyles and practices of the Jewish forefathers do not contain the anachronisms common to texts that might have been written well after the events that they describe. Kenneth Kitchen argues that archaeology and the Bible "match remarkably well" in their accurate portrayal of history. "In Genesis 37:28, for example, Joseph, a son of Jacob, is sold by his brothers into slavery for 20 silver shekels. That, notes Kitchen, matches precisely the going price of slaves in the region during the Old Kingdom (Egyptian) period as affirmed by documents recovered from the region that is now modern Syria. By the 8th century B.C.E., the price of slaves, as attested to in ancient Assyrian records, had risen steadily to 50 or 60 shekels, and 90 to 120 shekels during the Persian Empire in the fifth and fourth centuries B.C.E. If the story of Joseph had been dreamed up by a Jewish scribe in the sixth century, as many suggest, argues Kitchen, why isn't the price in Exodus also 90 to 100 shekels? It's more reasonable to assume that the biblical data reflect reality."[230]

But this kind of broad reasoning offers little material support and leaves Bible critics pointing to thinly veiled scriptural contrivances as evidence of misguided, event-based faith.

"Mainstream scholars are in the process of deleting ancient Israel from the history books. The entire period from Abraham the Patriarch in the -21st century (fundamentalist date) to the flowering of the divided kingdom in the -9th century (fundamentalist date) is found missing in the archaeological record . . ."[231] "As for the patriarchal stories, though they were valued for the light they cast on the beliefs and practices of the respective periods in which the various documents were written, their value as sources of information concerning Israel's prehistory was regarded as minimal if not nil. Abraham, Isaac and Jacob were commonly explained as eponymous ancestors of clans, or even as figures of myth, and their real existence was not infrequently denied. The patriarchal religion as depicted in Genesis was held to be a back-projection of later beliefs. In line with evolutionary theories abroad at the time, the actual

---

[230] *Is the Bible True*

[231] *C&AH*, vol. V, pt 2, July, 1983, p. 71

religion of Israel's nomadic ancestors was described as an animism or polydaemonism."[232]

Archaeology and the Historical Record

Is it possible that the dramatic stories of the forefathers in Genesis and those of the Book of Exodus -- stories believed by millions for millennia -- are completely absent from the historical record? If these stories are true, where would they appear in the chronology of the ancient world? And how has so distant a chronology – one either including or absent the biblical story -- been compiled at all?

The science against which all ancient history is measured is largely based on the archaeological "aging" of successive discoveries of various types of pottery. Digging deeper and deeper into the earth, experts have uncovered the remnants of older and older historic eras – as if the vertical chronology that they have traced is linear. It has traditionally been assumed that there can be no intermingling between the artifacts of one era and those of another above or below. To create this linear chronology, climactic, geologic and other changes that might mix the products of different eras have traditionally been ignored.

"A strong example of this ... (is) ... the famous flood level at ancient Ur[233] in southern Mesopotamia ... R. Milton gives this assessment of the massive scale of this flood as uncovered by Sir Leonard Woolley[234] at Ur: 'The extent of the Sumerian flood was very substantial: a deposit 8 feet thick, covering an area of some 400 miles long by 100 miles wide – a total of many billions of tons of material.' Common sense tells us that a flood of this magnitude must have affected other cities in the region as well. Yet archaeologists insist that the Ur flood did not reach even Eridu, a mere 20 kilometers from Ur, and situated at a level lower than Ur.

---

[232] *A History of Israel*, p. 66

[233] The ancient city of Ur was a city state in Mesopotamia, classically dated to 3800 BCE. It was under Sumerian and Akkadian rule. The area of Ur is active with archaeological digs and is thought of as the cradle of early civilization.

[234] Sir Leonard Woolley was a twentieth century British archaeologist known for his excavations at Ur. He is recognized as being among the first "modern" archaeologists. He discovered the tombs of many Sumerian royals.

Nor, according to Seton Lloyd,[235] is the Ur flood to be coordinated with other massive flood evidence elsewhere in the region, since . . . clean strata of water-borne sand or clay appeared in stratigraphical contexts which varied in time from the 'Ubaid period at Ur to the end of the Early Dynastic phase at Kish. At Farah (Shuruppak), however, a stratum of this sort occurs at the end of Early Dynastic I . . . ."[236]

If the classical linear approach to archaeology has failed here and elsewhere, then it needs to be replaced, perhaps "by more of a pond ripple effect stratigraphically . . . a strictly linear approach to stratigraphy can lead to severe anomalies and a failure to link contemporary peoples and events."[237] Kathleen Kenyon,[238] heralded for her discoveries at Jericho in the 1950s, became critical of the linear approach, growing to believe that many cultures existed side-by-side and that, as a result, older methods of dating produced inflated chronologies. We will soon see how shrinking these classically accepted chronologies brings ancient history into both accurate and practical focus.

FINDING ABRAHAM IN HISTORY

The ancient empire of Ebla in northern Syria was unearthed in the mid-1960s. In 1975 more than 15,000 tablets were recovered from this active archaeological site revealing a well-organized ancient civilization ruled by a detailed code of civil law. Conventionally dated to about 1000 years before the biblical time of Moses, the Ebla tablets reveal a language quite similar to biblical Hebrew, along with many common Hebrew names, including Abram, Israel, Michael, Esau, Saul and David. Among the texts were also references to cities known by the same names as those in the biblical account. A description of an

---

[235] Seton Lloyd was a twentieth century English archaeologist and academic who helped excavate sites in Egypt, Southern Turkey and Iraq. He wrote extensively on Mesopotamian history.

[236] "The Old Kingdom From Abraham to Hezekiah – A Historical and Stratigraphical Revision," p. 2,3 of 43

[237] Ibid, p. 3 of 43

[238] Dame Kathleen Kenyon, twentieth century archaeologist, was an expert on Neolithic culture in the Fertile Crescent. She is best known for her excavations at the ancient city of Jericho in the 1950s.

Ebrum who becomes king takes place at the same time as the adoption of the "ya" ending (a suffix referring to God) to names. Similarities in the language of Ebla and that of the Hebrews hints at Ebrum's identity as that of the biblical Eber, Abraham's ancestor. If so, the addition of the divine suffix to proper names is consistent with both the reputation of the biblical Eber and the later proselytizing of Abraham as described in the scriptural narrative and midrash.

The Ebla tablets refer to the five cities of the plain (including Sodom and Gomorrah) in the same order listed in Genesis. The tablets also contain a familiar account of creation.

Research by Herbert A. Storck correlates the kings lists of Mesopotamia and those of the Bible. He dates Abraham to the later part of the Ur III dynasty.[239] And Australian researcher Damien Mackey cites a quote from an official of Ashur in the *Cambridge Ancient History*[240] giving the name of the Sumerian ruler Amar-Sin. Mackey equates him with Amraphel of Shinar.[241] Both the Sin and El (Hebrew) endings refer to God. Even the conventional Mesopotamian dates for Amar-Sin and Abraham, his opponent in a biblical war correspond. The Ebla tablets also confirm the authenticity of names on the Assyrian kings lists. The biblical Tidal is accepted as a form of Tudkhalia thereby accounting for a second of the four kings described in the early Genesis war in which Abraham participated and liberated his nephew Lot.

The grandfather of Amar-Sin, Ur-Nammu, founder of the Ur III dynasty, was famous for both his conquests and for building massive ziggurats or towers to heaven. Could he have been Nimrod (thought to be the biblical, royal general contractor of the Tower of Babel)?[242] Mackey proposes that Enmaker, the builder of Uruk I (Genesis 10:10) was Enoch the son of Cain and that Gilgamesh was Lamech.

---

[239] *Proceedings of the 3rd Seminar of Catastrophism & Ancient History*, C&AH Press, Toronto, 1986, H.A. Storck, The Early Assyrian Kings List, The Genealogy of the Hammurapi Dynasty and the Greater Amorite Tradition, p. 43

[240] *Cambridge Ancient History*, vol. 1, pt. 2, 3rd Ed., p. 602

[241] Genesis Chapter 14, a contemporary of Abraham

[242] Nimrod was the Biblical son of Cush and the great grandson of Noah. He became the king of Shinar and is midrashically said to be the king at the time of the building of the Tower of Babel.

More correspondence between Abraham's biblical interaction with the local kings and archaeological evidence of their exploits comes from John Osgood.[243] "The passage in Genesis chapter 14, therefore allows us to conclude that in the days of Abraham there was a civilization in En-gedi . . . a civilization of Amorites, and that these were defeated by Chedorlaomer in his passage northward . . ." Excavations in the En-gedi area have found that the Chalcolithic period[244] of Palestine was the region's largest and most active settlement period. Osgood concluded that it must then correspond with the time of Abraham and the invasion of the four Mesopotamian kings (Genesis, Chapter 14).

Abraham's migration to Canaan is also biblically tied to early Egypt, as we see from his meeting and confrontation with an early pharaoh (Genesis, Chapter 20). And, the creation of nation states is tied to the migration caused by the destruction of Babel as described in Genesis. Historians and archaeologists can now tie that seemingly fictitious biblical story to actual events that occurred in the Early Bronze Age.

"The beginning of Early Bronze I in the late predynastic period of Egypt is tied in unmistakable fashion to Mesopotamian history for the period known as Jemdet Nasr. It is to be noted that, as in Egypt, so in Mesopotamia the Jemdet Nasr era marks the beginnings of dynastic history. Hence the point marks a widespread trend toward nationalism, as is to be expected following the Dispersion incident (Genesis, Chapter 11). Of this era, Piggot[245] wrote: '. . . We are now approaching so near to the recorded history and king lists of Mesopotamia that we can give an approximate date in years for the Jemdet Nasr – about 3000 BC – for it was followed by the period of the early dynasties.' The correlation of the beginning of Early Bronze I with the dispersion from Babel becomes reasonably complete if evidence is at hand to indicate that the short-lived Jemdet Nasr culture of Mesopotamia and other contemporary cultures became scattered over the area of the then known world. . . If one can

---

[243] *Ex nihilo Tech J.*, "The Times of Abraham," 1986, vol.2, issue 1, p. 77-87

[244] The Chalcolithic period is also known as the Copper Age, the early part of the Bronze age, immediately following the Neolithic period. It is believed to date to between 4500 and 3500 BCE.

[245] Stuart Piggot, the British archaeologist who died in 1996, developed a storied approach to archaeology.

free his thinking from the strangle-hold of popular opinion, the evidence becomes overwhelming that the beginning of Early Bronze I marks the point of the dispersion as recorded in the scriptural accounts. The magnitude of the migration of cultures at this point has been such as to call forth expressions of some astonishment on the part of scholars . . ."[246]

Famed archaeologist, William Albright[247] wrote: ". . . Towards the end of the fourth millennium there must have been an exceedingly intensive transfusion of culture going on in the Near and Middle East. Syria and Palestine naturally became the cultural intermediaries through which Mesopotamian influences streamed into Egypt in the period just before the first dynasty, as has been demonstrated particularly by Frankfort and Scharff."[248] And according to John Garstang:[249] ". . . In Palestine many great Canaanite cities have been shown by archaeological studies to date their origins from these times, such as Hazor, Taanak and Megiddo, on the north-eastern trade route, and Schechem, Beeroth and Jerusalem in the hill country to the south; and probably the same is true of most of the cities of the plains."[250]

Moreover, "the early bronze practice of multiple burials in large caves"[251] matches the form of burial chosen by Abraham, Isaac and Jacob in purchasing the cave of Machpelah from the Hitites.[252] Nor is it a coincidence that the ancient Egyptian, Hebrew and Mesopotamian languages have similar names for Egypt. The ancient Mesopotamian word for Egypt is Metzyr or a variant thereof, similar to the Hebrew Bible's Mitzrayim (the great, grandson of Noah). Ancient Egyptians

---

[246] "The Old Kingdom From Abraham to Hezekiah – A Historical and Stratigraphical Revision," p. 12 of 43 quoting Courville

[247] William F. Albright, died 1971, was an American archaeologist, biblical scholar, philologist and ceramics expert.

[248] Albright cited by Courville ("The Old Kingdom From Abraham to Hezekiah – A Historical and Stratigraphical Revision," p. 12 of 43)

[249] From 1930 through 1936 John Garstang, professor at England's Liverpool University, excavated Jericho, with particular attention to the biblical period.

[250] Garstang cited by Mackey, p. 13 of 43

[251] "The Old Kingdom From Abraham to Hezekiah – A Historical and Stratigraphical Revision," p. 33

[252] Genesis 23:9

referred to their land as Cham or Chan meaning black – much like the Hebrew Bible's Cham, Mitrayim's father.

Correcting a Corrupted Chronology

This research is suggestive, but is it persuasive? Without specific correlations and concurrences the conflict between traditional, academic Egyptology and biblical chronology – a conflict that has effectively identified the books of Genesis and Exodus as collections of fairy tales – cannot be resolved.

The accepted chronology of the ancient world rests on the work of two twentieth century historians, Eduard Meyer and James Breasted. Meyer, a turn-of-the-twentieth century German academic, established and verified ancient Egyptian dating through recorded sightings of the Dog Star, Serius (a technique known as Sothic dating).[253] Breasted, former faculty member at the University of Chicago is recognized as America's first Egyptologist. His textbook enshrined a list of ancient Egyptian kings in what the list's creator called "dynasties."

The commonly accepted Egyptian chronology, fixed by Professor Breasted a century ago, is based on the reconstructed kings list of Manetho.[254] Manetho, a priest and scribe living in Heliopolis two and a half centuries before the Common Era, recreated a lost Egyptian history. He listed 30 royal dynasties with 113 generations totaling 36,525 years in his *Aegyptiaca to Ptolemeios Philadelphos*. Whatever its historic value, his compilation was to an important degree an expression of pride designed to demonstrate to his Greek lords the great antiquity of his own, Egyptian people. (Similarly, the Chaldean, Berosus a priest of Belus, (285-247 B.C.E.) wrote in the *Chaldaika to Antiothos I* that his history of Chaldea was based on Babylonian astronomical records stretching back 473,000 years.) In translating Manetho, Professor Laurence Waddell, late 19th and early 20th century adventurer, explorer and Assyrianologist, suggests that the works of both Berosus and Manetho are primarily

---

[253] Eduard Meyer's dating of ancient Egypt is based on the 1460 year cycle of the Dog Star Sirius. Meyer's four Sothic dates have been proven to be inaccurate.

[254] Manetho was an Egyptian priest and historian who lived about 200 years before the Common Era. His *Aegyptiaca* establishes the classically known ancient Egyptian dynastic timeline and kings list.

expressions of the rivalry between Ptolemy and Antiochus who each proclaimed the civilizations over which they ruled the most ancient.

Such preposterous historical claims were commonly found in attempts by the conquering Greeks to reconstruct histories, the original records of which were recklessly destroyed during the conquests of Alexander the Great. Inaccuracies were exaggerated further by Eratosthenes (274-194 B.C.E.). In deriving the now accepted date for the Trojan Wars he assumed three generations, each lasting 100 or 120 years in order to arrive at a combined reign of 622 years for 17 Spartan kings. While this dating is now acknowledged as inaccurate by at least 300 years, it is still accepted and routinely applied by both historians and archaeologists to the histories of both Greece and associated cultures.

Manetho's Egyptian list comprises 75 kings from the first through the 19th of his designated dynasties. And, according to the chronologies of those lists, the most likely times for the Hebrew sojourn in and exodus from Egypt fall into time periods where no physical evidence of or historic testimony to their presence exists. For example, traditional Egyptology has no record of the biblical Joseph. None of the accomplishments credited to him in the Bible are identified in the generally accepted Egyptian timeline. And certainly none links a powerful, Joseph-like advisor to a great builder-pharaoh of what is now called Egypt's Old Kingdom.

However, since the mid-twentieth century confusion has reigned concerning the overall sequence, age and duration of the ancient Egyptian dynasties. It is now believed that many of the kings on this list actually reigned simultaneously, either as a consequence of name duplication, or concurrent reigns in the capital cities of Memphis in the north, and Thebes and Elephantine in the south. Donavan Courville reduces the timeline by as much as six centuries (*The Exodus Problem*). Peter James[255] stated that the Intermediate Period of Egypt (Dynasties 21-23) existed at the same time as other dynasties, reducing the classical dating of this period in his analysis by 250 years. In the introduction to that book, Professor Colin Renfrew of Cambridge University wrote, "The revolutionary suggestion is made here that the existing chronologies for that crucial phase in human history are in error by several centuries

---

[255] *Centuries of Darkness*

and that, in consequence, history will have to be rewritten. . . I feel that their critical analysis is right, and that a chronological revolution is on its way."

Historian Roger Henry wrote, "The problems in conventional chronology cannot be solved without challenging some of the basic assumptions. The numbered dynasties lists inspired by Manetho should be replaced. The numbers themselves are a problem, implying divisions where there are none, and family continuity where it does not exist."[256]

THE CASE FOR JOSEPH AND EGYPT'S OLD KINGDOM

Is there a place for Joseph in ancient Egyptian history after all? The Arab world has long held the biblical Joseph in high regard. According to Middle Eastern legend, Joseph was challenged in his later years by Pharaoh's younger advisors to demonstrate that his intellectual skills had not eroded. In response, he set about reclaiming part of a nearby desert for cultivation. In order to do so, he had water diverted hundreds of miles from the Nile to the area today known as el-Fayoum. The resulting lake, Birquet Qarun, is also called Bahr Yousef (the Sea of Joseph) by locals.

Imhotep was known to be the trusted and brilliant advisor to third dynasty pharaoh Djoser. Credited with the design of the Step Pyramid at Saqqara, he is called Asclepios by the Greeks for his scientific knowledge. An inscription on the Sehel stele (from the period of Ptolemy but copied from an earlier text) relates a 7 year famine during the reign of Djoser who sought Imhotep's advice and solution. This parallels the role that Joseph plays in the book of Genesis. Pharaoh's likeness also appears in Upper Egypt at Bet Khallaf next to someone called Zanakht (similar to the name given Joseph in the Bible by Pharaoh) for whom no tomb has ever been found. Professor Breasted and others discuss inscriptions describing Mentuhotep (one of the names of this famous, Joseph-like advisor) as "Vizier, Chief Judge, Overseer of the Double Granary, Chief Treasurer, Governor of the Royal Castle, Wearer of the Royal Seal, Chief over all the works of the King . . . Giver of good-sustaining . . . favorite of the king." In all, not dissimilar to the descriptions of Joseph

---

[256] *Synchronized Chronology*, p. 18

in Genesis. The name given by Pharaoh to Joseph, Zafanath-paneah (Gen. 41:45) is interpreted by A. Yahuda as "food, sustenance of the land is the living."

Then too, the Joseph of the Bible asked his children to insure that he not be buried in Egypt but that his remains be brought up to the Promised Land with the return of his people. As such, no burial place would be found for him in Egypt itself. Alan Gardiner suggests that the Step pyramid may have been intended for Imhotep (another of this advisor's names) whose tomb has never been found.

A coordination of the famines of ancient Egypt[257] and Donovan Courville's[258] comparison of the credentials of vizier Mentuhotep to those of the biblical Joseph converge when placing Mentuhotep in the 11th dynasty as does Professor William H. Shea. "In a word, our Mentuhotep, who was invested with several priestly dignities and who was Pharaoh's treasurer, appears as the alter ego of the king. When he arrived the great personages bowed down before him at the outer door of the royal palace."[259] ". . . Joseph was at once the 11th dynasty's Mentuhotep I, who unified Egypt; Mentuhotep II, who ruled for 51 years; Mentuhotep III, during whose reign we read – in the letters of the priest Hekanakht – of the onset of famine; Mentuhotep IV, about the state of whose reign N. Grimal has written: 'At this point the Turin Canon mentions seven empty years which correspond to the reign of Mentuhotep IV.'"[260]

The suffix "hotep" is common to Imhotep, Mentuhotep, and Ptahhotep, all of whom can be equated to Joseph. In fact, Ptahhotep belongs to the third and fifth dynasties according to differing accounts and is recorded as having lived 110 years just as did the biblical Joseph. Mackey notes the ancient Egyptian custom of assigning differing religious prefixes to a name according to geography, so that the Ptah might have been used in Memphis, and the Mentu at Thebes.

---

[257] *Famines in Early History of Egypt and Syro-Palestine*

[258] *The Exodus Problem and its Ramifications*

[259] *Egypt Under the Pharaohs*, p. 162

[260] "The Old Kingdom From Abraham to Hezekiah – A Historical and Stratigraphical Revision," conclusion

A case can also be made for Joseph having been the chief advisor to Unas, a 5th dynasty pharaoh who ruled from Elephantine. Genesis does say that Joseph ruled "all of Egypt," from the royal seats in both Memphis and Elephantine. In Saqqara at the pyramid of Unas an incomplete relief shows him attending to starving visitors (Joseph's brothers?), and inside appears the text "Unas will judge with Him-whose-name-is-hidden on the day of the slaying of the eldest" (perhaps a reference to the Exodus prophecy, known in the biblical narrative since the time of Abraham?).

Along with the confused and often redundant chronology of ancient Egypt, there are a number of other good reasons to believe that the very beginning of the Old Kingdom was the time of Joseph. Donovan Courville uses what he calls "eight lines of evidence"[261] to conclude that the first and third dynasties carried on simultaneously. He synchronizes the famine of Denephes of the first dynasty with the famine of Djoser of the third dynasty, whose advisor, Imhotep can be associated with Joseph as explained earlier.

PLACING THE EXODUS WHERE IT BELONGS

For 2000 years scholars have agreed that if the Exodus actually occurred, it must have taken place during what is known as ancient Egypt's New Kingdom. Though no experts claim direct evidence from Egyptian historical documents, many identify the Jews with the Hyksos[262] (Asian invaders of Egypt) and call their eventual expulsion the Exodus. Menetho writes that after their expulsion, the Hyksos went to Syria and built Jerusalem. Manetho also told a story of lepers from Egypt, who with the cooperation of the Jerusalemites later cruelly took over Egypt. Their chief adopted the name of Moses and led them to Palestine after their expulsion. Writing at the time of the Roman destruction of the second Temple, Josephus adopted Manetho's declaration that the Jews were the Hyksos. "Under any chronological system

---

[261] *The Exodus Problem and its Ramifications*, Vol. 1, p.170-176

[262] The Hyksos were thought to be an Asian tribe that invaded and conquered Egypt sometime after the sixth dynasty. Said to have reigned in Egypt for four centuries, they interrupted ethnic Egyptian rule before disappearing.

which can be reasonably advanced, the date of Israel's invasion (as the Habiru or Hyksos) and settlement falls within the period (1500 to 1100 BCE) when the country was ruled by Egypt as an essential portion of its Syrian empire."[263] If the Jews did conquer Palestine after leaving Egypt, then at least this four century time span accommodates the revised chronology suggested here. But if the Hyksos were not the Jews, then who were they? Let's first explore the case for the biblical existence of the Jews in Egypt. Only then will we be able to identify the Hyksos, their origin and their impact on Egyptian history.

Professor Eduard Mayer sought to support Manetho's chronology by utilizing Sothic dating to fix a firm date for the start of the 18th Egyptian dynasty (based on the recorded sighting of the Dog Star, Sirius --Sothis in Greek). But this evidence is flimsy, at best. The Ebers papyrus[264] on which the Sothic theory is based has more recently been shown to contain no actual calendar dates, and its lunar calculations seem to be in error. Further, no Sothic data at all is available for Egypt's Old Kingdom during which many different calendars were used.

Classical Egyptian chronology separates the Old Kingdom from the New Kingdom by about 700 years, and provides for two intermediate periods, one following each of these kingdoms. Courville has shown this to be a mistake, and in doing so unifies the Old and Middle Kingdoms on the one hand, and the First and Second Intermediate Periods on the other. Mackey further supports the validity of this revision by demonstrating that the technology used in the construction of the pyramids was not mentioned earlier than in Middle Kingdom texts.[265] In addition, Italian archaeologist, Emmanuel Anati proposed the Old Kingdom as the ideal time period for the Exodus, while Israeli archaeologist Rudolph Cohen suggested that it must have occurred during the 12th dynasty of the Middle Kingdom. If the Old and Middle Kingdoms were concurrent, as Courville, Mackey and others demonstrate, this difference of opinion disappears.

---

[263] *The Foundations of Bible History*, p.51

[264] The Ebers Papyrus was discovered in the 19th century and dates to 16th century B.C.E. Egypt

[265] See "The Old Kingdom From Abraham to Hezekiah – A Historical and Stratigraphical Revision"

The Elements of Jewish Exceptionalism

The list below is a recounting of Manetho's early Dynasties, those relevant to this discussion, as corroborated by later Greek sources.

Third Dynasty
- Necherophes, reigned for 28 years, also known as Sanakhte
- Tosorthios, reigned for 29 years, aka Djoser
- Tyreis, reigned for 7 years, aka Sekhemkhet
- Mesochris, reigned for 17 years, could be Khaba
- Suphis, reigned for 16 years
- Toserfasis, reigned for 19 years
- Aches, reigned for 42 years, may be Huni
- Sephouris, reigned for 30 years (or 48 years according to Eusebius)
- Kerferes, reigned for 26 years

Fourth Dynasty
- Soris, reigned for 28 years, aka Sneferu
- Suphis I, reigned for 63 years, aka Khufu
- Suphis II, reigned for 66 years, aka Khafre
- Menecheres, reigned for 63 years, aka Menkaure
- Ratoises, reigned for 25 years, aka Djedefre
- Bicheris, reigned for 22 years, aka Baka
- Sebercheres, reigned for 7 years, aka Shepseskaf
- Tamphtis, reigned for 9 (48 years according to Eusebius), aka Djedefptah/Thamphthis

Fifth Dynasty
- Usercheres, reigned for 28 years, aka Userkaf
- Sephres, reigned for 13 years, aka Sahure
- Nepherchres, reigned for 20 years, aka Neferirkare
- Sisires, reigned for 7 years, aka Shepscskare
- Cheres, reigned for 20 years, aka Neferefre
- Rathures, reigned for 44 years, aka Niuserre
- Mencheres, reigned for 9 years, aka Menkauthor
- Tencheres, reigned for 44 years, aka Djedkare
- Onnos, reigned for 33 years, aka Unas

Sixth Dynasty
- Othoes, reigned for 30 years, aka Teti
- Phios, reigned for 53 years, aka Pepi I
- Methusuphis, reigned for 7 years, aka Merenre I
- Phiops, reigned for 94 years, aka Pepi II
- Menthesupis, reigned for 1 year, aka Merenre II
- Nitocris (female), reigned for 12 years

The Seventh through the 11th Dynasties are known as the First Intermediate Period. Information is very obscure.

Seventh Dynasty
Julius Africanus refers to 70 kings reigning in 70 days. Eusebius records 5 kings in 75 days.

Eighth Dynasty
Africanus lists 27 kings for 146 years. Eusebius rcords 5 over 100 years.

Tenth Dynasty
All sources record 19 rulers based in Heracleopolis.

Eleventh Dynasty
Sixteen kings ruled from Thebes over 43 years.

Twelfth Dynasty
- Amenmenes, reigned 16 years, aka Amenemhat I
- Sesonchosis, reigned for 46 years, aka Senusret I
- Ammanemes, reigned for 38 years, aka Amenemhat II
- Sesostris, reigned for 48 years, aka Senusret II
- Lachares/Lamares, reigned for 8 years, aka Senusret III
- Ameres, reigned for 8 years, aka Amenemhat III
- Ammenemes, reigned for 8 years, aka Amenemhat IV
- Skemiophris, reigned for 4 years, aka Sobekneferu

The Thirteenth and Fourteenth Dynasties are known as the Second Intermediate Period. Few names are confirmed.

Thirteenth Dynasty
Sixty rulers spanned 453 years in all major sources.

Fourteenth Dynasty
Aficanus notes 76 rulers for 184 years, while Eusebius claims 76 rulers over 184 years in one source and 484 years in a different version of his history.

EVIDENCE OF THE JEWS IN ANCIENT EGYPT

Traditional historians see the Exodus, if it occurred at all, as little more than a minor, unrecorded, regional event. But according to the biblical account it was earth-moving. If the biblical account is true, then there must be material evidence of the presence of the Jews in Egypt, and the accepted timeline must be wrong.

In discussing textual evidence of the Exodus, Anati claims, "... the relevant texts do not date to the New Kingdom at all, but to the Old Kingdom ... the tribal social structures described in the Bible, the climatic changes and the ancient Egyptian literature ..."[266] While Anati does not overtly challenge the traditional dynastic dating, he acknowledges the strong evidence pointing to the end of the Old Kingdom as the time of the Exodus. Curator Emeritus of the Dead Sea Scrolls, Magen Broshi,[267] stated that archaeologists "didn't find a single piece of evidence backing the Israelites' supposed 40-year sojourn in the desert." But as Anati points out, none should be found in a New Kingdom context. Traditional dating has forced Archaeologists to assign their finds to the wrong time periods.

Given an adjusted timeline, parallels between the biblical and archaeological accounts abound. The Turin Papyrus lists the 6th dynasty king, Nefekare, also known as Pepi II, as having reigned for 94 years. According to the midrashic collection, *Sefer HaYashar*,[268] Melol, pharaoh of the oppression preceding the biblical exodus also reigned for 94 years.

---

[266] "The Old Kingdom From Abraham to Hezekiah – A Historical and Stratigraphical Revision," note 61

[267] Ibid, note 65

[268] *Sefer HaYashar* is an ancient Midrash whose introduction dates its origin prior to the destruction of the Second Temple. More common is the belief that the text was

The Turin papyrus indicated that the son of Pepi II ruled for just one year. Melol's son, Adikam, the pharaoh of the exodus, ruled for four years according to *Sefer HaYashar*, though the plagues occupied only the last year of his reign. While it was common for pharaohs to have many names, there was no hieroglyph for the sound of the letter "L" which, as a result, would have been pronounced like an "R." Melol might have been pronounced Meror, close to another of Pepi II's names, Merire.

According to the inscriptions at a tomb uncovered in Saqqara, Egypt, and later confirmed by Manetho, an insurrection took place during the 6th dynasty and a new king temporarily took the throne. This could have been a king from beyond Egypt "who did not know Joseph." Josephus also records that the crown had come into the possession of "another family" that became "abusive to the Israelites."

Evidence of natural cataclysmic destruction, including earthquakes on the Syrian coast also date back to the end of the Old Kingdom. According to archaeologist Claude Schaeffer and his excavations at Ras Shamra in the 1940s, these quakes reached Egypt and at least partly account for its end. Dr. Cohen, Deputy Director of the Israeli Antiquities Authority, suggests the Middle Kingdom's 12th Dynasty as the time period of the Exodus.[269] Like Courville, Down[270] and Velikovsky,[271] he points to the Ipuwer Papyrus as describing just those conditions.

The Ipuwer papyrus, discovered in Egypt at Saqqara in 1822, is the lament of an Egyptian priest. Translated by Alan Gardiner[272] in the early 20th century, the age of the papyrus is disputed. Ipuwer blames his pharaoh for the disaster that has befallen his country and demands action. The language of the papyrus also describes travails similar to those described in the biblical account of the ten plagues, which immediately preceded the destruction of Egypt. The sudden and complete

---

assembled from varied sources between the 9th and 12th centuries in Spain or southern Italy. It was first printed in Italy in 1553.

[269] "The Old Kingdom From Abraham to Hezekiah – A Historical and Stratigraphical Revision," note 66

[270] David Down is an Australian archaeologist who worked extensively in Israel and the Middle East

[271] Immanuel Velikovsky was a Russian-Jewish independent scholar and author of several books reinterpreting the events of ancient history.

[272] *The Admonitions of an Egyptian Sage*

collapse of the Old Kingdom in the time of Pepi II (according to most kings lists), who was followed only briefly by his second son, bears an uncanny resemblance to the midrashic, biblical history that names and describes the last two pharaohs before the Exodus. As noted earlier, Pepi II ruled for 94 years as did the pharaoh of the Bible. He would then be the new pharaoh "who did not know Joseph" after whose rule a madman presided only briefly over Egypt's destruction.

"It has been repeatedly pointed out the order of the plagues as described in the Book of Exodus is exactly the order of the annual discomforts caused by the climate and insects of Egypt under Turkish rule, and is largely the same today."[273] The Ipuwer papyrus offers a similar order and description.

First explained as a collection of proverbs and axioms followed by a chapter from a work of philosophy, this papyrus was at various times called a book of riddles, or a prophecy foretelling evil overtaking Egypt. Gardiner in 1909 claimed that it was a description of an actual national disaster.[274] He and others believe that the papyrus describes events from an earlier era: "The scribe used a manuscript a few centuries older."[275] Gardiner believed that the papyrus was describing one of two possible periods in ancient Egypt: "The one is the Dark Age that separates the sixth from the eleventh dynasty (the Old Kingdom from the Middle Kingdom); the other is the Hyksos period."[276] The revision of classical dating being discussed here satisfies both.

The papyrus reads: "Men ventured to rebel against the Uraeus (the royal emblem)" and magical spells associated with the serpent are revealed (6:6-7:6). Gold and jewels "are attached to the neck of female slaves." (3:2-3). This mirrors the Bible's promise to Abraham that his progeny would leave Egypt with wealth. It describes events that may have been the plagues themselves: "Plague is throughout the land. Blood is everywhere."(2:5) "The river is blood." (2:10) "Trees are destroyed." (4:14) "No fruit or herbs are found." (6:1) "The land is not light." (9:11) "No craftsmen work, the enemies of the land have spoiled

---

[273] *Ages in Chaos*, p. 14

[274] *Admonitions*, note to 1:8

[275] Ibid, page 3

[276] Ibid, p. 17

its crafts." (9:6). These phrases mirror the plagues of blood, hail, locusts and darkness.

In a reference to the biblical confrontation between the escaping slaves and the pursuing Egyptian forces at the Reed Sea,[277] the papyrus offers that the pharaoh commanding those forces disappeared under circumstances (7:1-2) "that have never happened before." According to the Bible, the Israelites were led to the Reed Sea by divinely provided pillars of fire and clouds – only to be confronted there by the pursuing forces of Egypt. Then, according to the biblical account, the sea split, the Jews safely crossed, and it closed on their Egyptian pursuers. Testimony to the veracity of that account may be found in a piece of carved black granite now housed in a museum in the town of Ismailia. Known as the El Arish Shrine (naot), the granite block has a hollowed center which once held an idol. The stone itself is covered in hieroglyphics telling the story of Pharaoh Thom, during whose last months Egypt and nature seemed to be in revolt. There was a period of intense darkness in Egypt itself: "The land was in great affliction. Evil fell on this earth . . . It was a great upheaval in the residence . . . Nobody left the palace during nine days, and during these nine days of upheaval there was such a tempest that neither the men nor the gods could see the faces of their next."[278]

The final battle between the Egyptian army and those described as "evil men" occurred at "the place of the whirlpool." The whirlpool was located at Pi-Kharoti where the pharaoh was thrown high into the air and never seen again. Not only does this description follow that of the book of Exodus, but the biblical place name of Pi-HaKhirot is almost identical. The Pharaoh Thom named here is listed in the Turin papyrus as Mem-tm-saf II. Recall also that the cities being built by Jewish slaves in the book of Exodus were called Pithom and Ramses.[279] Since Pi means "city of," the Jews would have been building the city of the reigning monarch, Pharaoh Thom (whose surviving son may have taken his name, as was sometimes done upon assumption of the throne) before perishing at the Reed Sea. The plague of darkness lasted three days according to the biblical description. The Midrash (Shmot Rabbah, Shir

---

[277] Exodus 14:21

[278] *Ages in Chaos*, p.40, quoting the translated text

[279] Exodus 1:11

The Elements of Jewish Exceptionalism

Hashirim Rabbah, and Sefer HaYashar) calls it a 7 day plague, while the naot at El Arish calls it nine days. All three accounts seem to describe the same event. The last day of that plague coincided with the arrival of the Jews at the Reed Sea – at a place called Pi-Kharoti in the El Arish naot or Pi-HaKhiroth as is recorded in the Bible.[280]

Egyptology generally recognizes that what it calls the Old Kingdom was a time of greatness for the empire. It is surprising, then, that this golden age came to so sudden and complete an end with the 6th dynasty. This abrupt end seems to be described in both the Ipuwer papyrus and the El Arish naot. It would be more than 400 years before Egypt would regain its standing as a world power.

THE JEWS AS SLAVES IN EGYPT

According to the Bible, the confrontation at Pi-Kharoti was an attempt on the part of the Egyptians to reclaim their liberated slave labor. If so, then there must be evidence that there was abundant slave labor available to Pharaoh at that time. "The Asiatic inhabitants of the country at this period must have been many times more numerous than has been generally supposed."[281] "It is apparent that the Asiatics were present in the town (Illahun/Kahun, described by Sir Flinders Petrie exactly as it was in the 12th dynasty) in some numbers, and this may have reflected the situation elsewhere in Egypt. . . these people were loosely classed by Egyptians as Asiatics, although their exact homeland in Syria or Palestine cannot be determined."[282] "Evidence is not lacking to indicate that these Asiatics became slaves."[283]

According to Breasted, 12th dynasty pharaohs carried out extensive building projects in the eastern delta region. This included the area known in the book of Genesis as Goshen. Abundant slave labor would have been indispensable to the empire at this time.

---

[280] Exodus 14:9

[281] *Cambridge Ancient History*

[282] *The Pyramid Builders of Ancient Egypt*, p. 190-191

[283] David Down, referring to the Brooklyn Papyrus describing Asiatic household slaves, which also list 77 slave names, 48 of which are common Hebrew/Semitic names

The same excavated town (Illahun) that testified to the abundance of Asiatic slaves may also testify to the Pharaoh's biblical decree that the male offspring of the Jewish slaves be slain. "Larger wooden boxes, probably used to store clothing and other possessions, were discovered underneath the floors of many houses at Kahun (Illahun). They contained babies sometimes buried two to three to a box, and aged only a few months at death. . . Interment of bodies at domestic sites was not an Egyptian custom, although such practices occurred in other areas of the ancient Near East."[284] And archaeologist David Rohl noted that graves in the Delta region from the same period had exceedingly large proportions of babies, 65% of all burials there being children 18 months of age or younger.[285]

Illahun also speaks to the sudden departure of Egypt's slave population (echoing the midnight exodus of the Hebrews as described in the Torah). "The quantity, range and type of articles of everyday use which were left behind in the houses may suggest that the departure was sudden and unpremeditated."[286] The continuous presence of dwellers in Illahun ends in the time of Khasekhemre-Neferhotep I of the 13th dynasty – a name quite close to that of Nefer-kare of the 6th dynasty.

Manetho describes the end of this dynasty as a period of anarchy with 70 kings having ruled in 70 days. The biblical narrative provides a rationale for this confusion. In the final biblical plague visited on Egypt, its first born sons were put to death, leaving no male heir to the Pharaoh -- who as mentioned above was documented to have disappeared at the Reed Sea.

Manetho's chronology identifies the last ruler of the 6th dynasty – the Old Kingdom -- as a woman named Nitocris. Listed on the Turin papyrus, she is believed to be the first woman Pharaoh. The Greek historian Herodotus recounts her reign as well. The sudden appearance of a woman in a line of ancient kings was typically a sign of disruption in the royal lineage.

Could Nitocris have been the surviving wife of the Pharaoh of the Exodus – the Pharaoh who mysteriously disappeared at the end of a

---

[284] *The Pyramid Builders*

[285] *Pharaohs and Kings*

[286] Rosalie David as quoted in Mackey note 70

failed campaign to reclaim his Asiatic slaves? Her reign was a brief one and coincided with the collapse of the kingdom and the empire.

THE CONVERGENCE OF BIBLICAL AND CONVENTIONAL CHRONOLOGY

Compressing the ancient historical record by hundreds of years is then little more than accounting for redundancies associated with concurrent dynasties or duplicate Egyptian monarchs. As a result of this revision, conventional dating and biblical dating converge, offering abundant physical evidence of the biblical Exodus, its precursors and aftermath. And there is further support as well.

Researcher Damian Mackey uses the example of Moses to link the 4th dynasty of the Old Kingdom and the 12th dynasty of the Middle Kingdom. If Moses lived in the 4th dynasty, then Joseph must have lived during the third dynasty. Similarly, if Moses lived during the 12th dynasty, then Joseph would have lived during the 11th dynasty. The 11th dynasty was characterized by a severe famine, a king-like second-in-command to pharaoh, the widespread presence of Asiatic slaves, and the unification (North and South) of the land of Egypt: exactly the circumstances of Joseph's tenure in Egypt!

Professor Anati links Moses to this time as well. He brings together these two dynastic periods through a recorded account of a high-ranking Egyptian official, Sinuhe. An ancient text records a story of "... Sinuhe, an officer of Pharaoh Amen-em-het I who lived in the harem and served the hereditary princess. It seems that he committed a violation of some sort, and when the Pharaoh died Sinuhe feared his successor." He then fled to "the land of Yaa near the desert" where he met a local chief, married his daughter and tended his flocks, before being called back to Egypt after the death of the Pharaoh.[287]

In Manetho's 12th dynasty list of multiple pharaohs, Amenemes I and III are likely the same person, and similarly, so are Sesostris I and III. The confluence of the 4th and 12th dynasties then fits chronologically. The traditional portrayals of Amenemes III also show him to have been

---

[287] "The Old Kingdom From Abraham to Hezekiah – A Historical and Stratigraphical Revision," note 32b

a brutal builder who used bricks and straw for building the Dahshur pyramid, for example[288]. French Egyptologist, Pierre Grimal noted an influx of foreign workers during the reign of this monarch.

"These (pharaohs) were followed by the ill-fated pharaoh of the Exodus, whose reign lasted only about a year, and whose first born son died in the plague (Exodus 12:29). The Old Kingdom's end came shortly after the 1-year reign of its last male ruler of the 6th dynasty, Merenre II, who can then be identified as above as the Pharaoh of the Exodus. His Middle Kingdom alter ego may well be Amenemes IV, Maat-en-ra (similar to Mer-en-re), whose mummy has never been found."[289] And since both the 6th and 12th dynasties ended with women as rulers, it is likely that, coincident with the Exodus account, no male heir remained.

The corroboration of biblical dating is of course controversial but may be lent additional support by the discovery of DNA markers that distinguish the priestly Jewish tribe. According to the biblical narrative, tribal status was conferred more than 3300 years ago. The integrity of that lineage has been preserved in modern times by, what is thought to be an unbroken, generational inheritance – an inheritance that would be manifest on the Y chromosome, a legacy of Aaron, the brother of Moses. A study in the mid-1990's (corroborated by subsequent studies) suggested that a large portion of today's priestly population carries genes in common, passed down through patrilineal descent from a single male ancestor. The results are not distinguished by North American, European, Mediterranean or Middle Eastern heritage. The study authors estimated that the common priestly ancestor lived about 106 generations ago, bridging the biblical Exodus and monarchy periods.[290]

Who were the Hyksos?

Following the biblical cataclysm comprising the plagues, the Exodus, the destruction of the Egyptian army at the Reed Sea, the disappearance of the Pharaoh and the collapse of the Egyptian empire, the Hebrews continued east into the Sinai and ultimately to Canaan /

---

[288] Ibid, note 49

[289] Ibid, note 55b

[290] *Nature*, Vol. 394, no. 138, "Origins of the Old Temple Priests"

Palestine. Egypt was left defenseless, ripe for domination by an invader hoping to take advantage of the general state of anarchy.

In addressing the convulsive and sudden fall of the Old Kingdom, Breasted wrote, "Exactly when and by whom the ruin was wrought is not now determinable, but the magnificent mortuary works of the greatest of the Old Kingdom monarchs fell victims to a carnival of destruction . . . The nation was totally disorganized."[291]

The catalysts for this destruction, the myriad Hebrew and other "eastern" slaves, were instrumental in changing the face of the region overnight. "The Asiatic inhabitants of the country at this period must have been many times more numerous than has generally been supposed. Whether or not this largely slave population could have played a part in hastening, or in paving the way for, the impending Hyksos domination is difficult to say."[292]

David Rohl offers some clarity here by describing the "gruesome scene" at Avaris where archaeologists found "burial pits into which the victims of some terrible disaster had been hurriedly cast. . . Bietak is convinced that we have here direct evidence of plague or some other sudden catastrophe at Avaris." And apparently an exodus from that city followed: ". . . site archaeology suggests that a large part of the remaining population of the town abandoned their homes and departed from Avaris en masse."[293] Their departure was followed by a foreign invasion [294] and then an occupation by Asiatic hordes who were not Egyptianized like the earlier inhabitants (the Hebrew slaves).

This growing cache of written and archaeological history tells us that the end of the Old Kingdom was the time of the biblical Exodus. And those disastrous circumstances "invited" the successful Hyksos invasion.

Who were the Hyksos, where did they come from and when did they enter Egypt?

---

[291] "The Old Kingdom From Abraham to Hezekiah – A Historical and Stratigraphical Revision," note 59

[292] *Cambridge Ancient History*, Vol.II, Part I, p.49

[293] "The Old Kingdom From Abraham to Hezekiah – A Historical and Stratigraphical Revision," note 71

[294] Ibid, note 73

Abu'l Faradj al-Isfahani (897-967) was an historian who is notable for having preserved ancient Arabic lyrics and poems in his *Kitab-Alaghaniy* (Book of Songs). He repeats a tradition that God sent plagues upon the tribe of Amalek[295] that forced it to leave Mecca and flee to the west. During its flight the tribe was overcome by ants, drought, famine, and intense clouds. Finally God led them to their native land only to send a deluge that drove them further westward. The Arab historian, Masudi (sometimes referred to as the Arab Herodotus, d. 956) also relates this tradition. He quotes the ancient poet el-Harit's statement that Mecca became a desert overnight. Eventually "the Amalekites invaded Egypt, the frontier of which they had already crossed, and started to ravage the country . . . to smash the objects of art, to ruin the monuments."[296] ". . . According to the Arabian traditions, the disaster which had smitten Egypt had also desolated their (Amalek) own homeland forcing the population to migrate to the relative safety and abundance of the Nile valley and delta."[297]

Before crossing into Egypt, the Amalekites were driven toward the shore of the Reed Sea by natural disasters comparable to those experienced by the Hebrews in Egypt. Simultaneously the liberated Israelites were moving east, toward Amalek. The biblical battle between the Israelites and Amalekites at Refidim occurred at that point,[298] after which they went their separate ways -- the Israelites into the Sinai desert and the Amalekites into Egypt. The Midrash actually states that the encounter took place in thick clouds, perhaps the same thick clouds that the Amalekites moved west to escape.

The Amalekites arrived to take control of an Egypt with neither an army nor royal family. The Hyksos occupied Egypt "without a battle" according to Menetho. They installed their own pharaohs and ruled for more than four centuries. Al-Samhudi (844-911) and Masudi both wrote of the Amalekites invading and destroying Egypt, while Persian historian al-Tabari (838-923) gave the genealogy of Amalekite pharaohs

---

[295] The tribe of Amelek is encountered and engaged by the Jews after their departure from Egypt. The Torah prescribes that the people of Israel both destroy Amalek and not forget them. Amalek is the grandson of Esau. The tribe is assumed to be nomadic and their historical origin is unknown.

[296] *The Meadows of Gold and Mines of Gems*

[297] *A Test of Time: The Bible from Myth to History*, p. 288

[298] Exodus 17:8

along with stories and legends. Four hundred years later, Amalek/Hyksos came up to attack Israel (Book of Judges, chapter 6), and in Samuel (chapter 30) David's troops find an Egyptian in the field who claims to be the servant of an Amalekite.

The Hyksos/Amalekite kings are listed by Manetho in the 14th to 17th dynasties, a period seen by him and subsequent historians as one of chaos, which, in written description, is nearly identical to the earlier period of darkness following the fall of the old kingdom. The end of this four century long Amalekite kingdom came with the defeat of their king Agog at the hands of Saul, the first king of Israel.   In   the book of Numbers (24:7), Bilaam, a pagan prophet, was hired to curse the Jews as they came within sight of the Promised Land on the plains of Moab. However, in looking down on Israel's encampment he instead blessed them saying "his king shall be higher than Agog." The Torah text continues with Bilaam saying, "When he looked on Amalek he took up his parable and said, Amalek is the first among the nations, but his end shall be that he perishes forever." (Numbers 24:20) Both the first and last Hyksos kings in the historical record of Egypt carry the name Apop, linguistically close to the biblical name Agag, that of both the invading king at the time of the Old Kingdom and that of the king defeated by Saul more than four centuries later. Neither the Amalekites nor the Hyksos were heard from again.

The Torah commands the Jews to wipe out Amalek, the same nation that we know today as the Hyksos. The prophet Bilaam testified to the pre-eminence of Amalek in antiquity. They swept out of the Arabian Peninsula into a decimated Egypt and conquered it, remaining its steward for four centuries. The recently emancipated Jews reached Refidim after leaving Egypt and crossed paths with Amalek on its way into Egypt. When Bilaam later blessed the Jews on the plains of Moab, he acknowledged that Amalek had already eclipsed the prominence of Egypt, but that its preeminence would end at the hands of the Jews. In fact, in another few generations, Amelek ceased to exist. The Jews established their kingdom in the land God gave them only to have it disintegrate in the wake of infighting and ultimately result in exile. They surrendered their divinely assigned leadership role and were dispersed among the nations.

Canonical literature accepts that diminished Jewish status. The text of the Book of Esther was adapted to reflect an eternal struggle with Amalek, even though Amelek had ceased to exist by that time and played no part in the narrative. The drama that was played out in Shushan 2400 years ago actually occurred at least five centuries after Saul destroyed Amalek and perhaps two centuries after its remnant was wiped out by King Hezekiah of Judah. No references to Amalek appear in the earliest Greek versions of the Book of Esther. The progenitors of the rabbinic movement may well have added references to Amalek to keep the message of Torah alive during the Hellenist and Roman periods, manifesting survivalist motives in lieu of the exceptionalist notion that might otherwise challenge civil authority and disrupt Jewish life in the growing Diaspora.

CHRONOLOGICAL AGREEMENT: EXODUS & CONQUEST

The altered dating described here offers additional support for the biblical conquest of the Holy Land by Joshua.

In her devotion to the classical methods of Egyptian dating, archaeologist Kathleen Kenyon forcefully disputed the aging of pottery found at Jericho. Her "proof" that these finds contradicted biblical chronology was accepted by historians and hailed by documentarians as proof of the fictional nature of scripture. But colleagues almost immediately began chipping away at her conclusions. Biblical archaeologist, Bryant Wood surmised that the destruction of the city walls of Jericho occurred just as was described in the biblical account. Wood also pointed out that wooden houses on the opposite side of town from the collapsed walls survived the destruction -- just as one would have expected of the poor side of town, the home of the story's heroine, the harlot Rachav. Subsequent excavations also uncovered granaries overflowing with produce. This is consistent with the timing of Joshua's attack (Passover harvest) and his command to his troops to take no spoil.

Excavations found that floors of the homes uncovered in ancient Hatzor, conquered by Joshua after Jericho (according to the biblical account) were charred from fire just as was described in the Book of

Joshua. And Kenyon herself declared that the decimation of cities in that vicinity at that time evidenced a common theme of intense burning.

In 1930 British archaeologist John Garstang noted that the double line of walls surrounding the city had been thrown down violently. The outer walls were tilted outward as if destroyed by an earthquake (re: the Bible story of the walls of Jericho tumbling down). Kenyon confirmed that "every town in Palestine that has so far been investigated shows this same break (complete destruction).[299] She also demonstrated that the last of Jericho's walls was built hastily and was probably not completely finished at the time of its destruction. This mirrors the biblical account of the city walls being feverishly reinforced by its ancient residents in anticipation of Israel's attack.

Work done by Yosef Garfinkel and his team at Khirbet Qeiyafa confirms the early establishment of the kingdom of David and links it to that of his son Solomon. Dating to the tenth and ninth centuries B.C.E., Khirbet Qeiyafa is identified with the biblical settlement called Sha'arayim -- mentioned in the David and Goliath narrative. The name Sha'arayim (meaning two gates) likely comes from the unique fortifications discovered by the archaeological team at Qeiyafa. A five line ostracon uncovered there is potentially the oldest extant Hebrew inscription, and some scholars interpret the text as referring to the birth of the Israelite monarchy. Garfinkel believes that the site was an administrative center of the new Davidic kingdom. The team wrote, "Khirbet Qeiyafa redefined the debate over the early kingdom of Judah. It is clear now that David's kingdom extended beyond Jerusalem, that fortified cities existed in strategic geopolitical locations and that there was an extensive civil administration capable of building cities. The inscription indicates that writing and literacy were present and that historical memories could have been documented and preserved for generations."[300]

The book of Numbers recounts that the Israelites camped in the plains of Moab, enduring a plague that killed 24,000 (Numbers, chapter 25). Later, in the book of Joshua, the Israelites crossed the Jordan River

---

[299] "The Old Kingdom From Abraham to Hezekiah – A Historical and Stratigraphical Revision," note 13

[300] *Bible History Daily*, www.biblicalarchaeology.org, 11/8/2013, Khirbet Qeiyafa and Tel Lachish Excavations Explore Early Kingdom of Judah

and camped in Gilgal before beginning the siege of Jericho. An earthquake destroyed the city's walls, the citizens of Jericho were massacred, and the city was burned to the ground.

"Precisely the same sequence of events is evidenced by archaeological finds from the final phase of Middle Bronze Age Jericho."[301] There is also evidence of a plague in and around Jericho immediately before the city's destruction. Shitiim is the site of the biblical plague in Numbers 25, and it was from there that Joshua sent spies to scout Jericho. In addition, scientific support for an earthquake (that would have been responsible for the walls of Jericho falling) having occurred at this time "comes from the tombs containing the multiple burials. Organic material contained in these tombs was found to be unusually well preserved. . . because a short while after the burials were made, natural gas containing methane and carbon dioxide entered the tombs and brought the process of biological destruction to an end."[302] The movement of the earth is the best explanation of the sudden release of this gas into the tombs.

Valinger[303] is yet another archaeologist who explains that the ". . . walls of Jericho fell flat outward from the city and show the effects of intense burning. These facts are uncontested and agree perfectly with the biblical narrative of the conquest of Jericho and Ai recorded in Joshua 6:1-8:29. . . Gibeon was one of the few cities not destroyed or abandoned at the end of EBIII (Early Bronze Age III) but . . . the site does show evidence of a radically new cultural influence at this time. This fits perfectly with the Joshua 9 narrative in which the Gibeonites deceitfully entered into a covenant with Israel in order to avoid conquest—and subsequently became subject to the invaders anyway."

This testimony to biblical accuracy had been ignored in order to accommodate long accepted Egyptian dating. To further support the chronological agreement between recent archaeological finds and biblical dating, Bimson adds that Egyptian scarabs found at Jericho with

---

[301] "The Old Kingdom From Abraham to Hezekiah – A Historical and Stratigraphical Revision," note 19, Dr. John J. Bimson of Trinity College in Great Britain

[302] Ibid.

[303] Ibid, note 10

the names of several pharaohs[304] make it reasonable to conclude that the walled city of Jericho was contemporaneous with the 6th and 12th dynasties of Egypt, offering further support for the unification of these dynastic periods as described earlier as well as the timing of the Exodus.

This corrected dating opens the door to a reconsideration of both the biblical and historical significance of both Jericho and its neighbor, Ai. Archaeological evidence shows that they were walled cities that were violently destroyed in the period immediately following the biblical Exodus. And archaeology's evidence of a massive cultural change throughout Palestine during that period points to the successful invasion by a new people.

Rudolph Cohen, former head of the Israeli Antiquities Authority, argues that this new culture appears first in the northern Sinai, then the southern Negev, across Transjordan and the Jordan River itself before reaching northern Palestine – just the path described for the Israelites in the Bible. This is corroborated by Willam Dever.[305] Cohen discovered pottery demonstrating that the area of Jericho had been occupied by Canaanites whose reign ended in destruction. They were then followed by a people (the Israelites) with very different pottery styles who settled in the hills rather than the valleys as had the Canaanites. William Albright[306] noted the Egyptian influence in both pottery and weapons among these new settlers. Cohen also remarked that these people routinely used granite and conch shells commonly found at that time in Egypt.

IN THE TIMES OF BIBLICAL KINGS

The many archaeological discoveries testifying to the veracity of the biblical narrative include the Assyrian records of Tiglath-Pileser[307]

---

[304] Ibid, note 21

[305] American archaeologist, William Dever was a professor of Near Eastern Archaeology and Anthropology at the University of Arizona.

[306] "The Old Kingdom From Abraham to Hezekiah – A Historical and Stratigraphical Revision," note 9

[307] An eighth century B.C.E. Assyrian king and conqueror of ancient Israel

which confirm his conquest of the House of Omri as described in the Book of Kings (2:15:29).

A lime plaster inscription found in 1967 at Deir Allah in Jordan displays the same writing technique used by Moses on the plains of Moab as described in Deuteronomy. This particular inscription offers a prophecy of destruction of the Moabites at the hands of the gods. And, just as found in the Book of Numbers, it is attributed to Bilaam, the son of Beor, a contemporary of Moses.

In a 1993 archaeological dig at Dan in the upper Galilee, Avraham Biran and Gila Cook discovered a flat, basalt stone with an Aramaic inscription. The basalt stone was part of a monument commemorating a military victory by the king of Damascus over two ancient enemies dating to nine centuries before the Common Era. The fragment identified one adversary as the "king of Israel," and the other as "the House of David." This discovery came as a shock to those convinced that the Davidic dynasty was little more than a myth. [308]

The Cairo Museum of Egypt displays a record of 19th Dynasty Egyptian conquests carved in stone. Discovered in 1896, it mentions the cities of Ashkelon, Gezer and Yanoam, and identifies Israel as a distinctive people. This is the oldest known reference to Israel outside of the Bible.

The Louvre in Paris houses the Moabite stone. Its inscription describes Omri the king of Israel who oppressed Moab, and it describes the successful rebellion of King Mesha. The text also refers to the tribe of Gad as having settled in Moab, just as is described in chapter 32 of the book of Numbers, both sources mentioning the settlement of Atarot.

An ancient obelisk from the time of the Syrian king Shalmanezer[309] depicts Israelites paying tribute to him. The text lists gifts from Jehu, five kings after Omri, and confirms both his rule and its chronology.

The Taylor Prism, held by the eponymous British colonel in Bagdad in the early 19th century makes direct reference to the biblical siege on King Hezekiah's Jerusalem. Neither this record nor wall carvings found in the palace at Nineveh, Sennacherib's capital, mention the conclusion of this military campaign which was, according to the Bible,

---

[308] *US News & World Report*, October, 1999

[309] King of Assyria who in 2 Kings 18:9 attacked Israel.

The Elements of Jewish Exceptionalism

Sennacherib's humiliating defeat.[310] Such documentation is typical of ancient recorded histories, recounting only victories – that is, with the exception of the catalogue of ancient Jewish history: the Bible.

In the mid-1970s a collection of clay seals was discovered in the hands of an antique dealer in Jerusalem and later in the nearby excavation of Yigal Shiloh. These seals were self-identified as the property of several Biblical personalities: Baruch ben Neria (servant of Jeremiah), Gemaria ben Shaphan (the scribe), Azariah ben Hilkiah (the high priest) and Jerameel (grandson of King Yehoiachim of Judah). Each plays a part in the pre-First Temple destruction drama unfolding in Kings II, chapter 22 and Jeremiah chapter 36 during the 18th year of the reign of king Yoshiahau of Judah. Hilkiah finds the book of the law (either the Five Books of Moses or the Book of Deuteronomy) and passes it on to Shaphan for confirmation. Shaphan reads it to the king prompting him to recommit his kingdom and people to the principles of the Bible while he initiates a campaign to wipe out idolatry. These finds confirm both the narrative and chronology of what was often assumed to be biblical fiction.

In 1952 a metal scroll was unearthed among the Dead Sea Scrolls, unique in that it was hammered out of pure copper. As its printing was in reverse, its primary text was intended to appear in the clay in which it was wrapped. Unlike the other scrolls, this one is an inventory of the holy treasures of the temple. It identifies 64 items and their secret locations – information intended to be hidden from the destroyers of Jerusalem. The scroll is on display in the Museum of Antiquities in Amman, Jordan. Yet, the original Dead Sea Scroll research team and those who followed dismissed the content of the scroll as fiction, the product of a Second Temple period renegade or forger. Unsuccessful attempts have been made to locate the items mentioned in the scroll, primarily the large caches of gold and silver.

Amateur archaeologist Vendyl Jones[311] correlated his research with *II Maccabees*, two marble tablets found in Jordan, the sources quoted by Rabbi Naftali Hertz in his *Emek HaMelekh* (1648), and indepen-

---

[310] 2 Chronicles 32

[311] A controversial Noahide scholar who directed archaeological digs for Biblical artifacts.

dent handwriting analysis to reach several unorthodox, speculative conclusions: The scroll was likely written in haste (at the time of the destruction of the First Temple) by the prophet Jeremiah and four colleagues, among them Shimor haLevi, who also wrote the marble tablets mentioned above and in the same bas style. Continued work at the site of the scroll's discovery by Jones and his colleagues has turned up what they claim to be two of the items mentioned in the scroll -- the anointing oil and the holy spices.[312] The authors of the scroll intended that its content be revealed at the arrival of the messiah, the rebuilding of the Temple and the reinstitution of the sacrificial service. While locations of the holy items on the list could be fictional, the discovery by Jones' team of two of the items mentioned would testify to the likelihood of the existence of the Exodus-era manna and staff (also in the copper scroll inventory) at the time of the First Temple.

It seems that it is a grave mistake to discount the Bible's value as both a spiritual guide and an accurate history. The misdirected research spawned by the arbitrarily and fancifully collected ancient Egyptian chronology presents a distorted narrative that has written the biblical Jewish experience out of history. "I suggest instead that we should treat Manetho's text as the work of an ancient historian," said David Nirenberg. "It is not the 'collective memory' of a society, but the creative redeployment of stories about the past in order to make sense of the present."[313] Rather than making sense of our present however, the ancient historian Manetho succeeded in corrupting the understanding of a theological and historical narrative whose followers were then subject to continual oppression and ridicule while being placed beyond the intellectual mainstream for millennia.

---

[312] When asked about the discoveries of holy spices and anointing oil, Lawrence H. Schiffman, academic and author of analyses of the Dead Sea Scrolls replied, "Those discoveries are of no importance. There is no ketoret discovery, scientists think it is animal dung. The oil is balsam, afarsimon, and has nothing to do with anointing that in any case didn't take place in second temple times . . . Either the stuff was buried and dug up or the Copper Scroll isn't historical. I assume the first but still have no proof." When the same was asked of Bruce Zuckerman, academic and expert on the photography and reproduction of the Dead Sea Scrolls, he replied, "My guess (and it is only a guess!) is that the document is not a fake."

[313] *Anti-Judaism*, p. 28

2. The Mother Tongue

How is the Hebrew language peculiar to the "chosen" Jewish role? The people and the language are forever linked by scripture. And like the people with whom it is identified, the language is exceptional – something more than just a system of communication. It is a database of both scientific and sociological information!

In the 1930s, linguists Edward Sapir and Benjamin Lee Whorf studied how languages vary. They concluded that speakers of different languages not only speak differently, but think differently. It has been demonstrated experimentally, for example, that people who speak languages that rely on absolute directions are better at keeping track of where they are than are either those who speak other languages or scientists themselves.

The rules of language actually train cognitive expertise. They can influence how quickly children learn the seemingly obvious. In 1983, three groups of children growing up immersed in Hebrew, English or Finnish were compared. Hebrew grammar identifies gender prolifically while Finnish does not and English has a little of both. Not surprisingly, those being raised to speak Hebrew recognized their own gender about a year earlier than the Finnish speakers with the English speakers in the middle. Findings also show that people who are bi-lingual see the world differently depending upon which language they are speaking. "Each (language) provides its own cognitive toolkit and encapsulates the knowledge and worldview developed over thousands of years within a culture. Each contains a way of perceiving, categorizing and making meaning in the world . . ."[314]

However, unlike other languages, Hebrew may have been created with imbedded technical data and wisdom.

The nature and purpose of scripture established Jewish exceptionalism, yet, it is the uniqueness of the language in which it is written that overwhelms the narrative itself. Jewish tradition explains that no letter, word or sentence in the Torah/Pentateuch is extraneous. The text has been composed with an efficiency that empowers each symbol

---

[314] *Scientific American*, February, 2011, p. 65,"How Language Shapes Thought," Lera Boroditsky

and sound with a largely unexplored innate meaning. Scripture was composed in a form that could well be appreciated by William Strunk, Jr. who wrote in his *Elements of Style* over a century ago, "A sentence should have no unnecessary lines and contain no unnecessary words, a paragraph no unnecessary sentences, for the same reason that a drawing should have no unnecessary lines and a machine no unnecessary parts."

The scope and power of biblical Hebrew has been denigrated in concert with the modern campaign against the divinity and historicity of the Jewish Bible. Simultaneously, the science of linguistics has developed universally accepted theories to explain the existence of the 6000 human languages. To the academic class, the seventy languages that were spawned by Hebrew at the biblical Babel in Mesopotamia thousands of years ago are part of a larger fairy tale. In reality, experts insist, clans developed several branches of language independently. There was no "mother tongue" as the Torah would have it, and even if there was, it certainly could not have been Hebrew.

Yet, until quite recently it was generally acknowledged that the Hebrew language was, in fact, the progenitor of all others. The story is told that Columbus brought a Hebrew speaker along on his trip to the New World, since he assumed that such an "ambassador" would be the most likely to be able to communicate with any new races encountered. Harvard and Yale included Hebrew among their initial course requirements and affixed it in their school mottos. What was arguably the first doctoral dissertation written in the New World was prepared at the Harvard Divinity School and it discussed Hebrew as the world's Mother Tongue. In the 18th century, The Continental Congress debated whether or not Hebrew should become the official language of the new nation.

In the first edition of his dictionary, Noah Webster cited Hebrew (Shemetic) as the source for many English words. For example, he identified the Hebrew word "yeled" as the source for lad, while the later Oxford English Dictionary simply states "origin unknown." As time passed and efforts to erase the originality of Hebrew progressed, the word "EMET" for truth in the motto of Harvard University was replaced by the Latin, "Veritas."

The academic pendulum now seems to be swinging back. The idea of a single, "first" language may not be so farfetched after all. "Today, scientists are leading a new revolution in understanding the roots of language. . . sifting through modern tongues for linguistic 'fossils' in the form of common words and grammatical structures . . . a few radical linguists have gone even further, claiming they have reconstructed pieces of the mother of them all: the original language spoken at the dawn of the human species."[315]

Academic scholars generally subscribe to the theory that among the many source languages that developed in different parts of the world, 13 language branches were spawned by one of the many mother tongues, this one linguistically identified as "proto-Indo-European." "While the existence of proto-Indo-European has been accepted among scholars for years, linguists have now begun to trace the lineage of languages back even further."[316] The geographic origin of this group of languages correlates to the Biblical geographic dispersion of the branches of the family of one of the sons of Noah, as identified in the book of Genesis.

Proto-Indo-European may be the language of the family tree of Noah's son Japhet. Genesis lists 14 branches of that family (not 13), 26 of Shem and 30 of Cham – 70 branches of the middle-eastern family of Noah in total. Later, the Bible describes the fracturing of language by families along these very lines.

It is now believed by many that we are all a part of a single family tree and that we have a common mitochondrial DNA ancestor who lived 200,000 years ago. The human hyoid bone is considered to be the foundation of speech, and it is now acknowledged that the oldest remains of a speaking human were found in what is now Israel. Why then is it not possible that a common ancestor may be the single source of human language?

"Proto-Hebrew"

The belief in the "monogenesis" of language -- that there is a single mother tongue from which all others arose – has of late regained favor in academic circles. Linguists Joseph Greenberg and Merritt Ruhlen,

---

[315] *US News & World Report*, Nov 5, 1990, William F. Allman, "The Mother Tongue," p. 60

[316] Ibid.

among others, have demonstrated the commonality of words and word structure across language.[317]

This is old news to biblical commentators who have long used this tool as a part of their exegesis -- commenting, for example on the biblical verses from Genesis Chapter 11 concerning the confounding of language at Babel. According to the Midrash, the words used there for "confound" and "confounded" are actually the mixing of language roots by God Himself.[318] The classical linguistic tool called Grimm's Law (whose use in Torah commentary preceded the Brothers Grimm by several centuries, and will be explained below) can be used to explain that action and result quite literally.

Edenics is the name coined by researcher Isaac Mozeson for the proto-Hebrew, mother tongue. The language comprises a proto-Semitic "Edenic" (from Eden) vocabulary in which each root letter has the "genes" for the wide diversity of the world's words.[319] Edenics is therefore the source code for all language and uses the accepted rules of linguistics to link words from different languages with their universal roots. Mozeson has shown this to be true across dozens of languages with over 50,000 words. He depicts how noun roots describe animals and other objects – Hebrew descriptions that are the building blocks of many other words. Mozeson's work demonstrates that the midrashic, biblical story of Adam choosing a name for each animal brought before him reflects the actual process of creating vocabulary from basic language.

The study of Edenics suggests that "there was a supernatural 'Big Bang' at the Tower of Babel that got the neurolinguistic diversity underway – for the purpose of enabling our multi-national history, and our multiplicity of perception... A case is made for the idea that humans no more evolved Edenic than they did chemistry or physics."[320] This analysis of language as a fundamental of nature is focused "on the external

---

[317] *Language in the Americas*, "On the Origin of Languages: Studies in Linguistic Taxonomy," and *The Origin of Language: Tracing the Evolution of the Mother Tongue*, NY, John Wiley & Sons, 1994.

[318] Midrash Tanchuma, Noach, Siman 28

[319] *The Origin of Speeches*, p. 3

[320] Ibid, p. 377

evidence, on the etymological data of Edenics. That is, how the uniquely versatile Aleph-Bet of a relatively small vocabulary – with only a few common linguistic shifts – can account for the vast diversity of global vocabulary."[321]

Unlike the identification of superfamilies of languages by modern linguists, the exploration and development of Edenics is focused on meaning, not the conventions of pronunciation or spelling. Students of Semitic vocabulary also understand that vowels are ignored in these languages – they are of use only in grammar. "But for other historical linguists to do similarly . . . would be too damn Semitic. It would mean giving up . . . that Sumerian might be Aryan, and ancient enough to scrap with the specter of Scripture."[322]

Mozeson's work shows that entire vocabularies from every major spoken language can be traced to proto-Hebrew. Examples include the words skeptic (Greek), Samurai (Japanese), and taboo (Polynesian). He shows how the biblical figure, Bilaam (hired to curse the Jews in Numbers 22-24) becomes the source of the word "blame." The Greek rendering of the biblical character Goliath (Kolios) spawns the words colossus, coliseum and colossal.

The desire of etymologists to draw from Greek and Latin only highlights their lack of biblical familiarity and their ignorance of Hebrew sources and roots. For example, sodium is believed by linguists to be derived from the Latin, soda. The "sound-alike" sod, from the Hebrew yesod means foundation, so the Hebrew source is dismissed out of hand. Instead, the English word (as well as in Latin) likely comes from the Hebrew for the city of Sodom on the salty Dead Sea shore. The Indo-European root for sparrow is sper (a generic for bird). The Hebrew equivalent is tsipor (its generic term for bird). While two languages can have similar sounding words, it's not likely that they would have the same meaning if they were unrelated.

The Oxford English Dictionary tells us that no direct Biblical connection can be traced for the word babble and that it is of unknown origin. Yet all etymological logic points directly to the biblical Bavel, the source of the language explosion itself.

---

[321] Ibid.

[322] Ibid.

Mozeson points to the Hebrew word "derekh" as an example that transcends language families and traces its Hebrew root meaning of "way," "manner," "journey" or "road" to words of the same meaning and root in 50 languages or dialects. Related words would then cover additional terms in many other languages. He shows that Grimm's Law, an accepted and conservative rule of linguistic change, designates seven English letter shifts that allow the discovery of Edenic roots for English words. And while the brothers Grimm popularized this tool, the biblical exegete, Rashi (11th century biblical and Talmudic commentator) actually explained this concept centuries earlier in his analysis of the biblical text.[323] A simple example would be the "b" and "v" sounds being interchangeable. Thus the German word uber meaning "over" would be the equivalent of the Hebrew word avar, also meaning over.[324]

Language is not the evolution of originally incomprehensible sounds from clans of cave men, but a natural science like physics or chemistry. A two letter root system is used with synonyms and antonyms to establish positive and negative meanings subject to root manipulation. Language is as intrinsic a part of the natural system as is any other science.

THE HEBREW LANGUAGE AS AN INFORMATION SYSTEM

But Hebrew is more than just the source of thousands of other languages. It is also a powerful data base in which the natural sciences are encoded.

Scientists have concluded that the universe is a mathematical concept and it was created with symmetries -- regularities in nature that allow for its mathematical description. The laws of nature, of physics, are consistent irrespective of space and time. Galileo called mathematics the language of science, and Einstein asked, "How is it possible that mathematics, a product of human thought that is independent of experience, fits so excellently the objects of physical reality?"

---

[323] Rashi's commentary on Leviticius 19:16: "All letters that begin from the same place in the mouth interchange."

[324] For a full explanation of Grimm's Law and allied linguistic tools as well as examples, see *Jewish Bible Quarterly*, Vol. 38, No. 1, 2010, "Could Pre-Hebrew be the Safa Ahat of Genesis 11:1," Isaac Elchanan Mozeson

These descriptions are expressions of the Formalist (smart people are responsible for the development of these fundamental sciences) school of thought.[325] Many others, including Roger Penrose and Kurt Godel, held the Platonist view, that mathematics exists in an immaterial realm (its existence preceded creation) and human beings can discover its truths. For example, these two approaches would argue whether quantum electrodynamics was invented or discovered to describe how light and matter interact. But, when physicists have used this mathematical construct to calculate the magnetic moment of the electron[326] the theoretical and experimental values seem to magically agree to within a few parts per trillion![327] Doesn't it seem that the real meaning or value of that scientific discipline is above and beyond the simple rules of mathematics?

What if the intrinsic, otherworldly nature of mathematics and, by extension, physics can be applied to language as well? Maybe Hebrew, to borrow the view of the Platonist, is just as fundamental to creation as are those disciplines, containing comparable secrets. On the other hand, if the Hebrew language is a creation of human-kind (in the language of the Formalist), then man deserves credit not only for this incredible, inclusive and versatile mother tongue, but also for its voluminous information content.

As noted above, until the Enlightenment, Hebrew was recognized by Western culture as much more than just another language.

Heinrich Cornelius Agrippa von Nettesheim (known as Agrippa) was an influential theologian, astrologer, magician and academic in early 16th century Germany. His *De Occulta Philosophia Libri Tres* was a summary of occult and magical thought that argued for the justification of Christian faith through a combination of natural, celestial and divine forces. ". . . Agrippa deals with the alphabet. In his view of the world, the order, the number and the shapes of letters are not accidental.

---

[325] Formalism in a specified discipline is concerned only with the rules of that discipline. There are no other truths that are beyond those rules.

[326] This is the electron magnetic dipole moment – the magnetic moment of an electron caused by its intrinsic properties of spin and electric charge. This comes from the rotation of an electrically charged body – the electron itself, which behaves something like a miniature bar magnet.

[327] *Scientific American*, August 2011, p. 81-83, "Why Math Works," Mario Livio,

They are not based on a convention that could easily have been different. Instead, the alphabet is related to the actual structure of the universe. This is in particular true of the Hebrew letters, which are the most sacred. The Hebrew alphabet consists of three parts: twelve letters are simple, seven letters are double and three letters are the 'mothers'. The simple letters correspond to the zodiacal signs, the double letters to the seven planets and the mothers to the three elements, earth, fire and water. The fourth element, the air, has a special position; it is the glue and spirit of the elements. In this way the Hebrew alphabet covers the entire universe and all kinds of relations between words and the world can be constructed. Wonderful mysteries concerning the past and the future can thus be drawn forth from the words that people use."[328]

Like hi-tech information systems, the Hebrew language, as Agrippa concluded centuries ago, comprises characters that are in themselves storehouses of embedded value and meaning. Each letter contains information that describes our world, our material reality. Edenics offers a systematic way to find information in the letter roots of Hebrew words. The information content of Hebrew also extends to the numerical value of Hebrew letters and words, as well as to the biblical context of those words. Just as has been described in the Midrashim of creation, the Hebrew language is the original tool box and database from which creation springs.

THE PERSONAL CONNECTION

Some time ago my daughter gave birth to a baby boy. That baby was carried for nearly nine months in her womb – her "rechem" in Hebrew. Rechem is also the root of the Hebrew word rachamim, which means mercy. The Biblical prophet Isaiah said[329] "Can a woman forget her suckling child, that she should not have compassion (m'rachem) on the son of her womb?" It is our understanding that the womb is the ultimate expression of human compassion. The woman steals from her own nutrition, from her own life force, to support the child growing within her.

---

[328] *Mathematics and the Divine: A Historical Study*, p. 26

[329] Isaiah, 49:15

Each letter in the Hebrew alphabet has a numerical value (aleph equals one, bet equals two, gimel equals three, and so on) – a necessity since Hebrew does not have a separate set of numeric characters. As a result, the numerical value of every Hebrew word can be easily calculated. The medical world tells us that human gestation is 266 days from fertilization or 280 days from the last menstrual period of the mother. That is an average of 273 days. The Hebrew word for pregnancy is Herayon.[330] The sum of the numerical values of the individual letters of the word Herayon is 271. This is the very number mentioned in the Midrash[331] as the length in days of human gestation. The word rechem has a numerical value of 248 and m'rachem (as used by the prophet Isaiah), 288. The average of both is 268. Due to concerns about the health and well-being of both my daughter and her fetus, the delivery of her baby was advanced by 21 days. Both mother and son emerged in good health.

Did the mathematics of the event and the word play of the Hebrew language somehow predict its outcome?

My daughter's Hebrew name has a numerical value of 220. Add the 21 days of prematurity to that and the result is 241. In order for the baby to make up for that 21 day deficit in time and get what one might call the full benefit of the mercy of his mother's womb, his name should have a value of 30, so that in combination with my daughter's 220, he will have benefited from a full measure of divine mercy (220+21+30=271). Unknowingly, or some might say, coincidently, she chose a name for her son with exactly that value! (In addition, the Hebrew spelling of my daughter's name is not traditional, and I cannot tell you how many times we were criticized for choosing that unusual spelling.)

Anecdotes like this often seem manipulated to offer religious inspiration. But scripture is full of better examples.

The Hebrew word *keri* appears seven times in chapter 26 of the Book of Leviticus and nowhere else in the Bible. English translations of the word include "casually" or "rebelliously," and to the individual who takes this posture toward God and His commands, seven kinds of punishment are promised. However, these common English translations

---

[330] Hoshea, 9:11

[331] Breishit Rabba, 20

are flawed. The word really refers to coincidence, or randomness. We know this from a reference in the Book of Samuel[332] where an Amalekite boy reports the death of King Saul to David (his rival and successor) and explains: "I happened by chance (nikro nikreti) upon mount Gilboa; behold Saul leaned upon his spear." The boy deliberately doubles the language of randomness to insure that his encounter with Saul could be understood as nothing other than pure chance.

In the book of Deuteronomy the people are told, "Remember what Amalek did to you on your way out of Egypt, when he occurred to you (asher karcha)."[333] For Amalek, "happening" upon the people of Israel escaping from Egypt was a chance occurrence and could not have been planned (see the earlier discussion of the Hyksos and Amalek to understand how the confrontation with the Children of Israel came about). [334]

In Egypt Moses addresses Pharaoh using the word nikrah more than once, as if to say that God suddenly and completely by chance appeared to the Jews. By contrast, Moses never uses that language when he addresses the Jews alone. While he minimizes the global significance of God's intervention to Pharaoh, he expects the Jews to understand the providential nature of God's actions on their behalf. Those actions are to be understood as anything but random. A form of the word is used again when God confronts the gentile prophet Bilaam.[335] Such language is never used when applied to Jewish prophets, but the Bible makes it clear that for non-Jews chance is very much in play.

Sidebar: There is an even more profound example of scientific and mathematical content of the Hebrew language that is demonstrated by the word *keri* And that is based on the fact that Keri has the same two letter Hebrew root as the word for "cold." Later, in addressing the Kabbalistic/Scientific aspect of Jewish exceptionalism, physics and consciousness will be discussed. It will be demonstrated there that cold and the scientific thermodynamic property known as entropy are conceptually identical. For the purposes of connecting quantum physical

---

[332] II Samuel, Chapter 1, verse 6

[333] Deuteronomy, Chapter 25, verse 17

[334] See the earlier history section to see how nicely this fits with the identification of Amalek as the Hyksos.

[335] Numbers, 23

properties and human free will, it will be explained that entropy and randomness are also equivalent. That discussion will identify the phenomenon known as Shannon Entropy[336] which, as defined earlier, is the mathematical equivalent in information theory of thermodynamic entropy[337]. Our upcoming scientific discussion will show (spoiler alert) that information is the "substance" of reality, and there comes a point at which it is difficult to distinguish between the subatomic particles of physics and information. Why does this matter? Because when trying to understand the concepts and logistics of creation the idea of "tohu vavohu"[338] in the Genesis story of creation will suddenly begin to make a lot more sense. Those Hebrew words in Genesis are variously translated as waste and void or chaos and desolation or formless and empty – each of which implies disorder (directly related to entropy) or cold as mentioned above.

Given this brief background, the fact that in Hebrew the same word refers to both randomness and cold (KR) is remarkable. Later in this text the concept of entropy will be discussed in more detail, but for now, suffice it to say that objects with lower temperatures generate larger entropy on absorbing heat than do hotter objects.

Shannon's entropy (as developed for the purpose of information systems) provides a measure of the amount of randomness that is contained in a specified statistical distribution. Increasing randomness -- in other words, increasing the uniformity in the distribution of probabilities -- results in higher entropy. More randomness implies increased entropy.

From this brief "sidebar," a basic understanding of creation along with the biblical use of the word keri is a little easier to understand: immediately after creation, as the universe expanded, its temperature fell and entropy increased. The lesson of the scriptural examples mentioned above and the "word/data mining" that was done is that turning away from God puts man in the hands of chance (*keri*). If chance and cold (khr) are inversely proportional, and statistical entropy grows with

---

[336] The statistical concept of entropy, developed by Shannon for information theory, measures the amount of randomness in a distribution of probabilities.

[337] This is the feature of physics developed by Boltzmann that is related to heat and temperature.

[338] Genesis 1:2

statistical randomness, we statistically increase the likelihood of an unplanned and undesired outcome in our lives by subjecting ourselves to "cold" or chance. This explains the punishments associated with this word in the section of the book of Leviticus in which the word *keri* is so prominent. Only a contemporary understanding of both scripture and the science involved makes it possible to draw the mechanics, operation and results of the exercise of free will directly from the language of scripture (stay tuned).

THE NATURAL SCIENCE OF THE HEBREW LANGUAGE

Associating mathematical concepts with the Divine is the substance of many ancient religious systems and was adopted by early Christianity. "St. Augustine was the first major Christian theologian who tried to show along Platonic lines that our knowledge of God is as certain as our knowledge of geometry. In St. Augustine's view mathematical knowledge is knowledge of an eternal abstract realm to which we have access by means of an inner light, Divine illumination. The existence of eternal truth in mathematics implies the existence of the idea of Divine Eternal Truth, which is an important ingredient of St. Augustine's proof of the existence of God and of the immortality of the soul."[339]

It seems that the Greeks, the Babylonians before them and pagans of all varieties associated numbers with the gods.

But biblical Hebrew embodied this natural characteristic from its inception. Biblical commentators knew for centuries that the numerical value of a Hebrew word (each Hebrew letter is also a number since there are no separate numerical characters) is directly related to the nature of that object or action. For example, the word "shanah" (meaning year) has a value of 355, essentially equivalent to the number of days in the lunar year (354.3671).

The same principle also applies to basic mathematical concepts. For example, the value of Pi can be calculated from this passage in I Kings, 7:23-26, "And he cast the pool, ten cubits from edge to edge, round, five cubits deep, and the perimeter surrounding it thirty cubits." At first glance the value of Pi ought to be 3 (30/10), however, in the text

---

[339] *Mathematics and the Divine: A Historical Study*, p.19

the word perimeter is written *kava* ending with the letter *heh*, and thus has a value of 111, but is traditionally pronounced *kav* (106). If we divide the two (111/106=1.0471698) and multiply by three we get 3.1415094...[340] Scripture describes Solomon calculating the value of Pi several hundred years before the Greeks did so.

We can now scientifically correlate many other, formerly obscure attributes of Hebrew words. The word for man or human being, *adam*, has a value of 45, the same as the number of human chromosomes (excluding the sex-determining chromosome). Similarly, the word for camel, *gamal*, has a value of 73, the number of chromosomes for a camel (excluding the sex-determinant). And in the case of the rat, or *choled* in Hebrew, the value is 42, corresponding to 21 pairs of chromosomes.

These relationships are not merely anecdotal or coincidental. Using linear regression analysis, Israeli professor and engineer Haim Shore[341] has demonstrated a relationship between many biblical words and the scientific or physical characteristics of those objects and actions that they represent. He demonstrates that in many cases, the numerical value of the words suggests that they were intentionally assigned to express actual, measurable physical characteristics. As is the case in the study of Edenics, each Hebrew word is the source of essential information, yet here the language becomes a vast database for our reality -- its limitation being our inability to decipher attributes that are beyond our current level of scientific knowledge or recognition.

From time to time man has waxed eloquently about the incredible beauty of nature and its innate intelligence. The Hebrew language manifests that intelligence. Imbedded in the language is a complete description of nature – an association of its content with every recognized natural science. Further revelation seems limited only by the pace and depth of our intellectual development. Jewish tradition depicts God as having used the Hebrew alphabet (which midrashically existed before physical creation) to create the world. As mankind becomes more technically

---

[340] As described by the Gaon of Vilna (Rabbi Eliyahu Kramer)in the 18th century, referenced here in *The Story of Pi*, Yitzchak Ginsburgh, Galeinai Publication Society, Jerusalem

[341] Professor in the Department of Industrial Engineering and Management, Ben Gurion University, Israel.

sophisticated and develops new tools with which to scientifically dissect nature, this traditional belief, thought for centuries to be nothing more than allegorical, may describe reality.

The Bible mentions five colors by name in Hebrew: red (adom), yellow (tzahov), green (yerakon), blue (techelet) and magenta (argaman). Using regression analysis Dr. Shore demonstrates that the numerical values of these Hebrew words are very closely correlated with the spectral wave frequency for these colors. Simply put, he plotted the numerical value of each word on the horizontal axis of a graph and the wave frequency on the vertical axis. The result was a straight line. This kind of analysis demonstrates that both sets of data are measuring the same thing, but on different scales. Shore has done this with 20 different sets of data, among them the relationship between the diameters of the earth, moon and sun.

Despite the limitation of our "primitive" scientific knowledge by Divine standards, we can still identify many Hebrew words that reveal important correlations. The Hebrew word for ears, *oznaim*, has the same root as the word for balance (*moznaim*). The discovery of the relationship of the ear to balance did not occur until 1874. The word for hand, *yad*, has a numerical value of 14, and the human fingers have 14 bones. Will we be able to examine correlations between the word *yad* and the hand's other natural attributes in the future? Dr. Shore takes the numerical values of the biblical metals, *zahav* (gold), *kesef* (silver), *bdil* (tin), *oferet* (lead), *nechushah* (copper) and *pladot* (iron) and correlates their numerical values to their atomic weights.

What we think of today as light appeared about 300,000 years after the Big Bang, after the period called "inflation," when helium and hydrogen atoms could finally coalesce in response to the cooling of the newly formed universe. The residue of that light is what we know today as the cosmic microwave background radiation.

Shore starts here and links the cosmological time scale with the biblical time scale of creation described in the book of Genesis. In order to demonstrate that both accounts are describing the exact same circumstances, he correlates cosmological age with biblical age (days) for several events: the creation of light, the formation of the first large scale celestial structures, the appearance of the sun and moon, the emergence of life on earth, and the creation of mankind (as detailed on an hour-by-hour basis on the sixth day by the Jewish oral tradition). His statistical analysis of the two creation accounts, one purely scientific and one purely biblical, shows their close correlation, and in statistical terms leaves only a 0.0165% possibility of such a result occurring by chance alone.[342] [343]

AND ABOUT THAT PREGNANCY AND BIRTH . . .

With the passage of time the technical value of Hebrew comes into sharper focus. Remember the earlier discussion of my daughter's pregnancy and birth? The word for pregnancy, *herayon*, has a numerical value of 271. Today we calculate the due date of the pregnant woman according to "Nagele's Rule" developed in the mid-19th century. In his study of patients, Dr. Nagele determined that a woman's due date is on average 280 days after the first day of the last menstrual cycle. In the mid-20th century it was determined that ovulation and conception take place 14 days after the first day of the last menstrual period. That defines a gestation of 266 days, plus or minus one day. The average of the two methods is 273 days. However, a 1990 study by Dr. Robert Mittendorf[344] determined that women who had never before given birth had an average gestation of 274 days, while women with at least one child had an average gestation of 269 days. The average of the two is 271.5 days – essentially an exact match for the numerical value of *herayon*!

---

[342] *Coincidences*, p. 283

[343] The notion of chance implies that quantum systems can act absolutely spontaneously, totally isolated from potential influences anywhere in the universe. The opposing position has it that all systems are continuously participating in an intricate network of causal interactions and connections at many levels. Individual quantum systems certainly behave unpredictably, but if they were not subject to any causal factors whatsoever, it would be difficult to understand why their collective behavior displays statistical regularities.

[344] *The Length of Uncomplicated Human Gestation*, Obstetrics and Gynecology, 1990; Mittendorf, R., Williams MA, Berkey CS, Cotter PF, 75:929-32. PMID 2342739. Quoted by Haim Shore in *Do Numerical Values of Biblical Hebrew Words Represent Major Physical Traits of Objects that the Words Stand For? Some Statistical Evidence*, Dept. of Industrial Engineering and Management, Ben Gurion University, December, 2011.

Each scientific revelation seems to advertise the incredible data content of the Hebrew language. The Hebrew word for blood is *dam*. Its numerical value is 44. The fact that there are four blood groups was discovered in the early 20th century. Shore demonstrates[345] that repeated word/letter values often identify a very significant characteristic of both the word in question and the object it represents. The letter/word *alef* in Hebrew has both the word value and the meaning of one and the repeated letter value 111. In biblical tradition, it signifies the oneness of God. The root of the word for firstborn in Hebrew is *bekor* which has a numerical value of 222. The firstborn is Biblically entitled to a double portion of the inheritance. Snow is *sheleg* with a value of 333, possibly representing the three states of water.

Returning to blood, we have learned that science has provided additional detail concerning its composition. We now see how the Hebrew word "dam" accounts for that information as well. Shore cites a source noting that 44 is also the percentage of the cellular, non-plasma content of human blood (hematocrit). Experts have estimated that the hematocrit percentage varies from 42% to 50% for males and from 35% to 47% for women. Standard medical guides put the average at 44.

The Hebrew language is a unique vessel that embodies a seemingly infinite amount of data describing nature, the bulk of which we are only now beginning to discover through scientific inquiry. According to this description, Hebrew is the data set from which nature arose and continues to be managed. The essential composition of creation seems to be pure information. What is the source of that information? How is it best used?

It may be that the world is a poorer place for the denigration and minimization of the Hebrew language.

In the wake of political and existential turmoil, and later, subjugation to their host nations, Jewish leadership became complicit in minimizing the divine nature of the language of scripture. The translation and dissemination of Hebrew scripture in Greek[346] both degraded the

---

[345] *Coincidences*, p. 146

[346] Mishna, Megillah, 1:8, Shimon ben Gamliel states that Greek is the only language other than Hebrew in which one can write a scroll of the Torah. The Jerusalem Talmud in tractate Megillah 71C says that the sages checked and found Greek to be

status of scripture, and relinquished access to the insight that was to be drawn from the divine language itself. Educational leaders of the Jewish community demonstrated to the gentile world that Hebrew was a language not appreciably different from thousands of others. It did not take long for it to become diminished in the eyes of the Jews themselves. From the all-inclusive database of creation, it was reduced to a common tool of communication, and then to little more than a scriptural language. With that diminution of status, Jews themselves became just another nation with a national tongue associated primarily with a homeland. Exile reduced Hebrew to a vestigial language akin to Latin, stripped of its profound content.

3. The Kabbalistic and Scientific Tradition

Is there something uniquely Jewish about the "mystical" Kabbalistic system?

After all, mysticism is also integral to both Christian and Islamic theology. Yet this Jewish tradition is so ancient that it is understood to have been given to Moses at Mount Sinai along with the written Torah. As such, it was an essential part of the oral tradition which was passed down from generation to generation until late in the Second Temple period. At that point religious leaders were convinced that even the most learned among them could no longer understand the true meaning of these teachings. It was excised from the standard oral canon for the safety and protection of the less learned. Only after the destruction of the Second Temple, shortly after the advent of the Common Era did some of the "tanaaim" (rabbis of the Mishnaic period) reestablish a connection with the mystical tradition. They later incorporated it into the Talmud (most notably in the Mishna, Chagigah, Chapter 2). Yet, the rabbis of the Talmudic period forbade the public teaching of the mystical tradition, and its transmission was limited to just one student at a time. The danger associated with this field of study was highlighted

---

the only language that retains the actual meaning of the Hebrew words. But later, in the extracanonical tractate Sofrim 1:67, this is contradicted, citing there the inability of any language to accurately express the meaning of Hebrew words. It notes that the day that the Torah was translated into Greek was as bad a day for the Jewish people as was that of the sin of the Golden Calf in the wilderness.

in the Babylonian Talmud's story of four sages who simultaneously experienced a mystical attachment to the Divine. "Four men entered PARDES (PRDS) – Ben Azzai, Ben Zoma, Acher (Elisha ben Avuyah), and Akiba. Ben Azzai looked and died; Ben Zoma looked and went mad; Acher destroyed the plants (became a heretic); Akiba entered in peace and departed in peace."[347] Mystical knowledge and study was so dangerous, that in even this group of four great sages, only one emerged unscathed by its study.

This story is also the source of the four varieties of textual interpretation credited to biblical commentators. The four levels on which the Torah is understood are found in the acronym PRDS, the Hebrew word for the "orchard" of mysticism into which the four sages ventured. The most basic of these four levels is called "Peshat," that is, the plain meaning of the words themselves. Next, there is "Remez," a way to take hints from the text and develop allegorical meanings. A third level of understanding is that of "Derash," a body of exegesis that builds on comparisons across texts and stories.

The fourth and most esoteric level of understanding is that of "Sod," the hidden, inner meaning of text as studied and expressed in Kabbalah. Here, the ties between science and religion are found at the most fundamental levels.

THE SCIENTIFIC AND SPIRITUAL FUNDAMENTAL OF UNITY

"Recent computer simulations by Rob Crain of the Swinburne University of Technology in Melbourne, Benjamin Oppenheimer of Leiden University in the Netherlands and their collaborators suggest that up to half of the baryons currently locked into galaxies in the local universe have cycled through the intergalactic medium at least once and often many times. The baryons that make up your body have participated in this cycle for nearly 14 billion years; the matter within your fingernail could have formed in stars in other galaxies and then spent billions of years exiled in intergalactic space before coming to rest in

---

[347] Babylonian Talmud, Tractate Chagigah 14b

our solar system. You are just an ephemeral phase, a brief host, to this rare substance we call 'normal'."[348]

Asian traditions (Buddhism, Taoism, Hinduism, particularly Advaita Hinduism and Theravada Buddhism) speak of the *undifferentiated* unity of the entire universe. This is quite different from the scientific description of unity. The wholeness that physics examines is *highly differentiated* and structured (much like the classical Kabbalistic template with its well defined order and imagery). It is subject to strict constraints, symmetry principles, and laws of conservation. In Einstein's relativity theory, space, time, matter, and energy are all unified. But, there are specific rules that govern a transformation from matter to energy, for example. In the structureless unity of the Far Eastern religions all distinctions have been eliminated. Those systems do not account for structured interaction and behavior that is typical of systems, such as those in biology.[349]

"Philosophers and scientists have dreamed of a unified view of all nature since the scientific revolution of the seventeenth century," says Steven Weinberg. They want to find direct relationships among the basic forces of nature that can explain existence. Interestingly, divine, biblical revelation, has always had its own "unified theory." Are these two versions of unity related? Can they both be expressions of a single truth?

Basic to the belief system of Judaism is the statement from the Mosaic code declaring that God is a unity – He is One. And that same definition is applied to the universe, an agglomeration of all kinds of seemingly very different things. So, what is a unity and how does that term apply to God?

"But, however great the variety presented to you both by nature and by history and by your own life. . . this is the doing of One God" wrote 19th century German rabbi Samson Rafael Hirsch. " . . . One God everywhere and in everything. Everything comes from this One God both in heaven and on earth, and everything therefore conforms to one design, is part of one all-wise plan."[350]

---

[348] *Scientific American*, May, 2011, p.53, "The Lost Galaxies," James E. Geach,

[349] *Religion and Science*, p. 189

[350] *Horeb*, S.R. Hirsch, Unity of G-d, Paragraph 6, p. 5

There is a religious notion that God has made each individual an integral part of the overall unity of creation. Circumstances testify to the fact that energy, matter and sentient beings work together in ways in which they are unaware. The unity of creation is manifest in these relationships and in the information that makes these connections possible.

The individual participants and actors in our universe – the moving parts, so to speak -- seem to understand their interdependence. "The question of information comes up time and time again. How does information get around so fast on a sub-atomic level? How does matter know? Matter seems to know . . . Iodine 125 has a half-life of 65 days . . . you can predict with absolute precision that every 65 days half of the radioactivity is lost . . . although half of the atoms will be gone in 65 days; any individual atom may last for one second, for a millisecond or for many years. There is no way to predict . . . Nonetheless, somehow they figure it out amongst themselves, in agreement. During the 65 day period, the atoms nudge each other, 'now you go.' 'No, I'll go.' 'No, you go.' And it always comes out precisely. It's not 64, it's not 63, it's not 66. It's always exactly 65 days. How do the individual atoms know? . . . The implication . . . is that there is knowledge, intrinsic and inherited, in physical existence. Although we consider physical matter to be inanimate, really it is sensitive -- something alive and filled with information. It knows itself, and it knows what's going on within itself. This is basically a mystical concept -- that there is knowledge which is innate within matter. This is a natural consequence of observations made in quantum physics and it's unavoidable."[351]

THE UNITY OF GOD AND THE UNITY OF NATURE

At the same time that the Bible is speaking to the unity or "oneness" of God, nature is speaking of its own unity.

The biblical statement, "Hear oh Israel, the Lord our God, the Lord is one"[352] is the bedrock affirmation of monotheism. Its internalization is the acceptance of the belief in a single divine Creator and Administrator. But is "oneness" the same as unity?

---

[351] *Sparks of the Hidden Light*, p. 49, quoting physicist Dr. James Brawer

[352] Deuteronomy, Chapter 6

The medieval rabbi and biblical philosopher Bachya ibn Pekuda wrote, "The idea that we should form in our minds of oneness is of absolute uniqueness and solitariness; that which has no association or comparison whatsoever, is totally devoid of plurality and number, and has been neither combined with anything nor separated from anything."[353] ". . . the true nature of His essence is a Oneness that intrinsically contains and encompasses everything that can be considered perfection. All perfection therefore exists in God, not as something added on to His existence, but as an integral part of His intrinsic identity, whose essence includes all type of perfection . . . Admittedly, this is something far beyond the grasp of our understanding and imagination, and there hardly exists a way to express it and put it into words."[354]

Conceptualizing the unity of God begins with the acceptance of the fundamental notion that He is the creator and that everything else is created. From that starting point comes an understanding that an absolute unity implies perfection (whatever that is), and that perfection implies absolute simplicity. For this reason, many religious commentators refuse to even discuss any individual attributes or characteristics of God since a perfect unity cannot possibly be subdivided. We violate this approach from time to time in order to make His existence a little more understandable to us. But, as a practical matter, doing so ascribes a kind of plurality to Him, which, of course, contradicts the principle of His absolute simplicity.

Whether we talk about God's Will or God's Intellect or God's Mercy, we are talking about God as a unit. In principle these features or attributes do not exist independently. Each is intrinsic to Him and is as much a unified representation of Him as is any other. It is also incumbent upon us to try to grasp the idea that a unity in this sense is not subject to change either, because whatever does change has a different status before and after the change, hence, plurality.

"Even in truth from His side, God, from the point of view of His essence, fills everything without barriers and without any differentiation or change of places," reads Nefesh Hachaim (by the 19th century rabbi and kabbalist, Chaim Velozhin). "Everything is a simple unity just as

---

[353] *Duties of the Heart*, Chovot Halevavot, Shaar hayichud, perek/chapter 7

[354] *The Way of God*, Derech Hashem, 1:1:5

before creation. Even so, we are not able or permitted to enter into a contemplation to know and comprehend how the Master fills everything with a simple unity!"[355]

But the world of science has explored this topic and arrived at its own definition. Here the word "universe" implies a certain "relative" unity -- a thread of commonality among disparate but related parts.[356] Science began with the universe and defined units that seemed to be only remotely related. It divided itself into discrete disciplines -- physics, chemistry, biology. Over time relationships were discovered between the discrete parts. More recently the parts were shown to be tied to a greater, organic whole. But, does that greater whole somehow mirror the simple unity that we've called God? A millennia ago *Chovot Halevavot* (Bachya ibn Pequda) spoke of this distinction[357] and differentiated between the two, one absolute and the other relative. Absolute unity is ascribed to God, "it is the origin of all that is plural," while relative unity is ascribed to His creation (the universe).

A discussion about the universe and the relationship of its parts to the whole concerns relative unity. While the General Theory of Relativity is really a description of gravity, its name reflects the fact that nothing in our universe can act in absolute independence of all else -- and that is a classically biblical concept. It weaves its way through discussions of free will and providence, and is most prominent in Kabbalistic texts and discourse.

Must the implications of these two theories, the "relative unity" described by *Chovot Halevavot* that is attributed to the universe here, and general relativity as described by Einstein and his successors, be mutually exclusive? The language and mindset of each is so similar to

---

[355] Nefesh HaChaim, Shaar 3, Perek 6. This prohibition to explore God's essence is contradicted in the works of some others: Chovot Halevavot (10th Cheshbon in Cheshbon HaNefesh), Rambam (Chapter 51, Vol. III of the Guide), Ramchal (Mesilat Yesharim, Chapter 21 of the Dialogue Version, Derech Hashem 4:2 on the Shema, Mesilat Yesharim, Chapter 25, Adir Bamarom on Yirah V'Yichud)

[356] "Space and time are inseparable, mass is a form of energy, and gravity and acceleration are indistinguishable. . . Matter is, if you will, a wrinkle in the elastic matrix of spacetime. Instead of separate enduring things, externally related to each other, we have a unified flux of interacting events." *Religion and Science*, p.180

[357] *Duties of the Heart*, Chovot Halevavot, Shaar Hayichud, perek 8

that of the other that the comparison between the two is clearly more than stylistic.

That Einstein's Relativity is called a theory tells us that there are many ways to look at the interrelated parts of the universe. Similarly, Talmudic sages offered that there are 70 "faces" to the Torah – i.e., numerous ways (theories) to interpret the words of God as recorded in scripture and equally as many systems of understanding them. Every belief-tradition is manifested in a different way, depicting cultural and historic influences, philosophical approaches and hidden, mystical meanings. No single biblical belief-tradition can claim its explanation as the truth to the exclusion of all others. Similarly, the theory of relativity is but one of many theories that seem to legitimately portray physical relationships.

". . . progress . . . has been made recently in finding dualities or correspondences between apparently different theories of physics," wrote Stephen Hawking. "These correspondences are a strong indication that there is a complete unified theory of physics, but they also suggest that it may not be possible to express this theory in a single fundamental formulation. Instead, we may have to use different reflections of the underlying theory in different situations . . . this would be a revolution in our view of the unification of the laws of science."[358] Substitute the word "Bible" for "physics" and this statement might have been ascribed to a professor of theology at a major university. In the cases of both physics and scripture, a more accurate and understandable picture of the truth is achieved by borrowing from all of the supportable theories available.

Natural law testifies to the concept of unity in creation. For example, it has been known for some time that many characteristics of living creatures, including their life spans, pulse rates, and rate of energy consumption vary according to body size. Animals all seem to have about a billion heartbeats in a lifetime. Smaller animals just use them up more quickly. There is a precise mathematical principle obeyed here called quarter-power scaling. And researchers have demonstrated the extension of this "law" to the plant kingdom as well. Scaling manifests itself in the subatomic world and shows up in Heisenberg's Uncertainty

---

[358] *A Brief History of Time*, p. viii

Principle too!³⁵⁹ It is scientifically and naturally fascinating that everything seems scale dependent – it is woven into the fabric of the universe. Population density, the average number of offspring produced, and time until reproduction are dependent on body size scaled to quarter powers. And these same laws apply to cities and corporations.³⁶⁰ It has been empirically demonstrated that these laws do not apply discriminately, but universally.³⁶¹

---

[359] Chaos Theory represents that in chaotic systems the smallest uncertainty in initial conditions can lead to huge uncertainties in the prediction of subsequent behavior. "If God acted by making infinitesimal changes in chaotic systems it would be undetectable scientifically; divine action could not be proved, but not disproved either." *Religion and Science*, p. 183

[360] "Of Mice and Elephants: A Matter of Scale," Jan 12, 1999, George Johnson, University of CA at Santa Barbara High Energy Physics, and "CEOs for Cities," 7/14/07, Conversations, blog, The Living City

[361] Biological scaling describes ways that life spans, pulse rates, energy consumption and other characteristics change according to body size. As animals get bigger, their pulse rates slow and life spans grow. The number of heartbeats during an average life is roughly the same for most species – about one billion (although humans get about twice that according to the San Jose State University Animal Longevity and Scale – likely due to both our attention to health and to industrial advances). This is one of many natural phenomena that changes according to a precise mathematical principle called quarter-power scaling. A cat has about 100 times more mass than a mouse and lives about 100 to the one-quarter power (about 3 times) longer. The heartbeat of a species is related to the mass of the animal to the minus one-quarter power. Therefore the cat's heart beats a third as fast as that of the mouse. (This, according to a study done by Geoffrey West, Santa Fe Institute, a physicist at Los Alamos National Lab, and two biologists at the University of New Mexico -- Jim Brown and Brian Enquist). For metabolic rates the relationship to mass is based on a three-quarter power calculation called Kleiber's Law. The rate at which a cat burns energy is 31.6 times greater than that of a mouse – 100 to the ¾ power. This relationship holds across the animal kingdom, through single-cell organisms, and possibly within cells (mitochondria) themselves. "Everything around us is scale dependent. It's woven into the fabric of the universe." (West)

Scaling cannot be linear because body mass increases along three dimensions. If a man is a million times more massive than an ant he will be only 1,000,000 to the 2/3 power stronger, about 10,000 times – so, he could lift hundreds of pounds, not thousands. Sorry, Spiderman!

"It is truly amazing because life is easily the most complex of complex systems, but in spite of this it has this absurdly simple scaling law. Something universal is going on." (West)

Population density, the average number of offspring, and the time until reproduction are all dependent on body size scaled to quarter-powers. There is even an analog to Kleiber's law in the plant world. It is assumed that scaling relationships arise from

The Elements of Jewish Exceptionalism

Quantum physics gives us a picture of the innate unity of materiality at the most elementary level. The sub-atomic property called "entanglement" describes the constant relationship (entangling) between all discrete aspects of creation. Every single part of creation is continually interacting with others as if each knows itself to be a part of a greater whole. And it is likely that this elemental relationship can be extrapolated to our life-size, macro level as well. ". . . you, too, retain a quantum bond with your loved ones that endures no matter how far apart you may be. If that sounds hopelessly romantic, the flip side is that particles are incurably promiscuous, hooking up with every other particle they meet. So you also retain a quantum bond with every loser who ever bumped into you on the street and every air molecule that ever brushed your skin. The bonds you want are overwhelmed by those you don't. Entanglement foils entanglement, a process known as decoherence."[362] Work done at universities in Russia and Poland as well as at M.I.T. suggests that entanglement can play a role in large, biological systems and not only at the microscopic level. When two of us make "a connection," there is a lot more to it than meets the eye.

This idea can be ramped up to a gigantic level – the universal unity from which everything else emerged. Both biblical scripture and general relativity recognize that there is a starting point for creation and that

---

the mathematical nature of the networks that both animals and trees use to both transport nutrients to their cells and carry away their wastes. The human circulatory system and the roots and branches of a tree look remarkably similar. In these systems each small part is a fractal network: a replica of the whole. For example, if we magnify the network of blood vessels in a hand, the image resembles one of a complete circulatory system. And if we assume efficiency in area, when a branch splits, the cross sectional areas of the daughter branches must add up to that of the parent – this way blood or sap continues to move at the same rate through the organism. A three-quarter power scaling is produced between the metabolic rate and the body mass of a plant. To make the calculations work in animals we try to visualize a heart pumping liquid through these fractal networks. It seems that evolution has overcome the natural limitations of simple geometric scaling by developing very efficient fractal-like webs. (Or was that the original design in keeping with the holographic nature of creation?) Mitochondria inside the cytoplasm and even the respiratory components inside the mitochondria are arranged in fractal-like networks. West is working to see if river systems (which look like circulatory systems) have similar scaling laws. ("Of Mice and Elephants: A Matter of Scale")

[362] *Scientific American*, "Easy Go, Easy Come," George Musser, Nov., 2009, p. 25

this is when time began. Time, space and matter were created simultaneously [Which, by the way, is not universally believed by biblical commentators. The early biblical commentator Abraham Ibn Ezra for example, believed in the Islamic dualism of the time, positing that God created the world from pre-existing matter which, like God, had existed forever![363]] General relativity offers that the universe is not static, and that it is in fact expanding. Stephen Hawking notes that "An expanding universe does not preclude a creator, but it does place limits on when he might have carried out his job!"[364] Quantum thought takes creation a step further in demonstrating that there was an original unity that still persists. But the mystical, religious equivalent was part of man's thinking well before the fundamentals of physics and cosmology were discovered.

THE VOCABULARIES OF JEWISH MYSTICISM AND SCIENCE

It may seem counterintuitive, but the Jewish Kabbalistic tradition—an allegorical pageant of spiritual imagery -- explains creation in much the same way as does theoretical physics. The vocabularies are distinct, but both describe objects, forces and actions thrust into existence at its very dawn. In a very real sense, each offers a glimpse into the "mind of God," seeing the divine, or as many prefer, the uniquely random set of circumstances that resulted in existence as we know it.

Tradition ties the introduction of Kabbalah to the giving of the Torah at Mount Sinai. Its spiritual development is presented in the Zohar, claimed as a tanaaic (early first century rabbinic) work, written largely in Aramaic, but not publicly revealed until its "discovery" in Spain in 1284 by Moses deLeon. The Jewish, mystical school of thought was known by different names in early rabbinic literature, and later as *chochmat emet* (the wisdom of truth) by Spanish biblical commentator Nachmanides in the early 13th century. According to tradition, mystical knowledge was transmitted orally by the forefathers to the prophets and sages, and ultimately to the people, so that by the 10th century before the Common Era, residents of Jerusalem had woven it into their daily

---

[363] Ibn Ezra on Genesis 1:1, and *Philosophies of Judaism*, p. 120

[364] *A Brief History of Time*, p. 9

practice. It is said that the Sanhedrin restricted the transmission of Kabbalah after the destruction of the First Temple, deliberately hiding this knowledge from subsequent generations. As mentioned earlier, the Babylonian Talmud makes a point of further limiting the spread of this knowledge in Tractate Chagigah.

Medieval Christianity renewed interest in and adopted Kabbalistic thought, and many *rishonim* (Jewish exegetes of the Middle Ages) relied on it, at least in part, in their biblical commentaries. Sixteenth century Rabbi Isaac Luria (known as the Ari or Arizal) espoused a "mystical" construct that formalized and simultaneously revolutionized the understanding of Kabbalah. His vision of this "hidden" Torah was built on the language of the 10 sephirot, a framework believed to have been first presented over 2000 years earlier in Sefer Yetzira (The Book of Creation), a book of unknown authorship, credited by some to the forefather Abraham. To this the Ari added new terminology and concepts that broadened Kabbalah and touched upon every aspect of creation, life and the afterlife.

The Ari presented more than just concepts. He proposed a nearly tangible structure to the spiritual world -- a structure of shapes, designations and "near-physical" combinations of spiritual-physical alter egos whose unions and actions were a precondition for existence and action in this, the physical world.

To his 18th century follower, Rabbi Moshe Chaim Luzzato (Ramchal), the constructs of the Ari were the reality of the hidden, spiritual world of truth. "Kabbalah is nothing but an explanation of how the holy Emanator ordered the laws of guiding power, how the holy Emanator causes and guides all matters of His world with great wisdom."[365] A century later, the Vilna Gaon (Rabbi Eliyahu Kramer) and his student, Rabbi Chaim Velozhin, understood the Ari's conceptual framework as a *mashal*, a parable that allows us to conceptualize and understand the otherwise incomprehensible. Ramchal himself said that "Kabbalah has its own language . . . in minute detail, Kabbalistic language explains every stage of the world's creation . . . No science can develop without

---

[365] *Song of the Soul*, p. 10, citing Ramchal, Milchemet Moshe

its own vocabulary. Scientists could not communicate without a shared language. They can relate to concepts only by developing terms."[366]

While philosophers and halakhists (experts in religious law) have developed their own ways of understanding God and His creation, only the Kabbalist envisions a pre-existing world -- a world that continues to exist side-by-side with ours in an immaterial dimension. "Just as God in His infinite ability created physical realities that we can see, so too did He create realities superior to those, unfelt and unsensed by us."[367]

The commonality of these two systems – that of physical science and the other of spiritual "science" -- is striking. Each can be considered a parable describing a reality that is otherwise incomprehensible to human kind. And each reveals progressively greater levels of both mystery and understanding as new details are discovered.

A strong case can be made that science and Kabbalah describe the same reality. It may be that the language of Kabbalah is simply a forerunner, a collection of images and language that preceded the incomprehensible technicalities of theoretical and quantum physics. If these two systems of thought are really representations of the same reality, then what language and concepts do they share?

Kabbalah offers a structured way of processing the enormity and complexity of creation. That complexity is presented in the form of familiar images and human characteristics to which we can relate. We are to understand something of the unknowable, that the infinite and inexplicable essence of God, His unlimited will (called the *Ein Sof* in this mystical terminology), is the source of not only everything we recognize as physical reality, but of the immaterial reality that preceded it.

A brief primer on the Kabbalistic notion of creation follows. Parallels with familiar scientific concepts should be apparent as the description unfolds.

In order to get the creative ball rolling, the infinite God/Creator had to limit Himself, in terms of our understanding, in order to make creation (which is something that is imperfect, with limitations) possible. And since we conceptualize God as a perfect unity, He is, as we

---

[366] Ibid, p. 11

[367] Ramchal, Maimar HaIkarim

discussed above, the equivalent of His own will. He had to limit His will in a process known in Kabbalah as *Tzimzum*. By doing this, He "carved out" a place for creation. This place, the *chalal* (meaning empty space), contained nothing but the original divine "light"—the instrument that God used to carve out that space. Once that light was withdrawn all that remained was a self-contained space with the original light of creation surrounding it. The residue of the creative light that was left behind in this space is called the *Reshimu*. And even though the space contained only a remnant of that original light, that remnant was sufficient to support the coming creative process.

Next, God had to fill this empty space with something.

Into this otherwise empty space (a space now containing an equally distributed but powerful remnant of "original light") was introduced a single ray of that supernal light -- a kind of laser beam from the *Ein Sof* called the *Kav*. As the *Kav* passed through the void it established ten vessels or mechanisms whose purpose would be to administer the remainder of the creative process. These *sephirot* represent the ten Kabbalistic dimensions of creation. They are vessels that on the one hand hold part of the creative light and on the other became the roots of our physical reality.

The Kabbalistic theory tells us that every aspect of our physical reality has a genesis and a parallel in the spiritual world. In addition, there are cause and effect relationships between the activities of our physical world and their spiritual alter egos. Had the early masters of Kabbalah known of astrophysics, atomic and sub-atomic worlds, they would very likely have seen the commonality of the two constructs.

Modern science tells us that our reality began with a physical "singularity" called the Big Bang. Theoretical physicists admit that speculation on their part about what preceded the Big Bang is meaningless. So how do we correlate the spiritual and the physical if God's creative process began as we've described, with His self-limitation in order to create a "space" for creation as we know it? We can start by drawing some simple parallels.

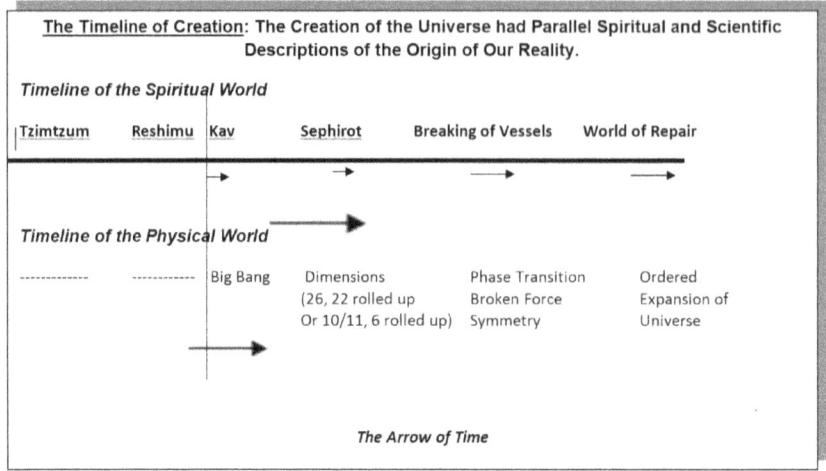

A classic mystical religious commentary asks, "One of the basic theological problems is concerned with the seeming enigma of reconciling God with the universe; how can there be a transition from the Infinite to the finite, from pure Intelligence to matter, from absolute Unity or Oneness to multifariousness?"[368] Let's work on that.

The Mishna in tractate Avot (5:1) says, "The world was created by means of ten utterances." Both the Kabbalistic and scientific systems have a view on that, and there are remarkable parallels.

"In the beginning, everything was simple light and we call that *Ein Sof*," says the Ari. "There was no empty space, rather everything was *Ein Sof*. When it arose in His will to emanate emanations, He limited Himself in one central point by way of this light. He limited himself on all surrounding sides, so that all around the sides remained empty space."[369] The process of creation began with an unlimited God carving out an infinitesimally small space in which His essence was limited. That self-limitation was confined to a single point.

Five hundred years later, theoretical physics arrived at the same conclusion. After overcoming both Aristotle's assumption that the world and matter have coexisted forever, and Kant's deduction that there are equal grounds to believe either of these theses, science settled on the

---

[368] Introductions to Tanya, Part IV

[369] Ari, Eitz Chaim

Big Bang. In 1929 Hubble discovered that the universe was expanding from the explosion of a single point of infinite density 15 billion years earlier -- The Ari's *Tzimtzum*. That explosion was not only the beginning of matter, but the beginning of time.

"This secret, the *reshimu*, is the place where all reality exists."[370] Into this void, now filled with the residue or radiation of God's divine light, came a purposefully directed "beam" of that light, what the Ari called the *kav*.

"After this space (*reshimu*) was defined, the Creator, Blessed be He, desired to emanate the emanations, the creations, the formations, and the present lowly world. In order to emanate all the worlds and the places that were made for them, the Creator Blessed be He, decreed that a *kav* be diffused. It would be like a thin pipe extending from the light of the *Ein Sof* which surrounds it, and continuing into the *chalal* (void). Via the *kav*, the light from the surrounding infinity descended into the space in proper measure."[371]

The *kav* gave order to the *reshimu* by filling it with the presence of ten "vessels" -- the ten (or 11 as well be explained below) *sephirot*. These vessels represent the dimensions of our reality. It is as if the Big Bang was triggered by the *kav*, leading immediately from infinite and simple order to a rule-based, growing disorder.[372] The action of the *kav* completed the transition from the infinity of the Creator to the finite nature of creation.

---

[370] Ibid.

[371] *Song of the Soul*, p. 170, quoting Eitz Chaim

[372] The second law of thermodynamics explains that a physical system dies when it gets to the point of maximum disorder: it reaches a point at which it can handle no additional information. This information overload is also known as thermal death. Every system must evolve toward its maximum disorder. Consequently, the second law is telling us that life, as a system, must also eventually end. Physicists tell us that the entropy of a closed system always increases. Since both physics and information theory have the concept of entropy, entropy can be seen as quantifying the information content of a closed system. And, the second law tells us that we cannot convert one form of energy into another without losing some energy to the environment in the process. When we burn fuel, for example, some of the energy is converted into motion and some is lost as heat. This inefficiency increases disorder/entropy, and contributes to Global Warming, for example.

But we don't live in a ten dimensional reality, do we? The String Theory of quantum physics explores that possibility. And a further connection between these systems is drawn from the Kabbalistic allusion to the dimensions of reality being represented in the ancient, holy combination of spices (incense) that are called *qetoret*. These were used in the priestly service in the Holy Temple in Jerusalem.

Twentieth century rabbi, commentator and physicist Aryeh Kaplan explained this relationship according to the understanding of the Ari: "The incense consisted of ten spices or perfumes with good fragrances, and an eleventh spice, chelbenah-galbanum, with a vile odor. These spices were each ground separately and then blended together to be used in the Temple. Since these eleven spices parallel the eleven *sefirot* of the Universe of *Tohu* (chaos), they are therefore said to represent the complete rectification of evil. This is indicated by the addition of the eleventh spice, chelbenah-galbanum, which alludes to the elevation of evil back in to the realm of the holy."[373] For now, the idea of evil can be identified as an imperfection in creation that needs fixing.

According to the traditional Kabbalistic understanding of creation, the shattering of the vessels -- the breaking of the *sefirot* of the Universe of *Tohu* (chaos) -- occurred when God allowed a bit of His infinite light to shine into those vessels, destroying them. The Midrash Bereshit Rabbah 3:7 records that, "In the beginning, the Holy One created worlds and destroyed them, created worlds and destroyed them." In the immaterial scene that we are describing here, seven of the *sefirot* shattered and four did not. Therefore, God's action seems to have left four of the vessels from which reality emerged undamaged. These correspond to the dimensions of space-time. And this also seems to correspond to the four incense spices that are actually mentioned in the text of the Torah. The names of the other 7 spices that comprise the unique combination used in the holy rites come to us from the oral tradition. It is from these lower seven *sefirot* or dimensions that the nuts and bolts of our physical universe seem to have emerged. Perhaps these seven dimensions, unlike space-time, are hidden from us -- rolled up, in the language of String Theory.

---

[373] *Innerspace*, p. 86

The Elements of Jewish Exceptionalism

Here Judaism also strays from the dualist idea of good and evil having been created simultaneously but separately. Judaism and its Kabbalah understand that evil is a direct consequence of the shattering of the vessels, that is, a distortion of good. Free will allows man to behave in a way that can "elevate" evil to again become part of the good. Had good and evil been separate creations, this freedom would not exist.

THE NEXUS OF THE PHYSICAL AND SPIRITUAL

The Big Bang was our physical reality's equivalent of Kabbalah's "Shattering of the Vessels." In the first moments of physical creation, the big bang generated a temperature so high that matter could not exist . . . so in a very real way, physical creation was not actually physical at all. Many believe that creation was something of an abstraction, a mathematical construct equivalent to the metaphysical shattering of the last of the *sephirot* -- a bridge between the immaterial and the physical. Time began with that "bang," and Kabbalah calls the emanation of the *kav* (see above) at that point the stimulus or trigger for creation.

Physical existence is defined by the beginning of matter and time in the context of our four dimensional reality (space-time). But as a practical matter, why should our reality be four dimensional?

"Two space dimensions do not seem to be enough to allow for the development of complicated beings like us," says Stephen Hawking. "For example, two-dimensional animals living on a one-dimensional earth would have to climb over each other in order to get past each other . . . similarly, it is difficult to see how there could be any circulation of the blood in a two-dimensional creature. There would also be problems with more than three space dimensions. The gravitational force between two bodies would decrease more rapidly with distance than it does in three dimensions . . . the orbits of planets, like the earth around the sun would be unstable . . . the least disturbance from a circular orbit would result in the earth spiraling away from or into the sun. We would either freeze or be burned up. In fact, the same behavior of gravity with distance in more than three space dimensions means that the sun would not be able to exist in a stable state with pressure balancing gravity. It would either fall apart or it would collapse to form a black hole. On a smaller scale . . . we could not have atoms as we know them."[374]

The dimensional orientation necessary for the orderly creation and physical development of the universe is mirrored in the creation of those Kabbalistic vessels, the *sephirot*. They are the engines of creation and the transition between an infinite Creator (who exists outside of the system) and His finite creations (that are part of the system that He created – the universe). They are the receptacles and transmitters of divine will and power.

The Ramchal wrote in the 18th century, "Even though we do not understand how they came into being from nothing, we do know that the *sephirot* brought (all the material creations) into being."[375]

Our reality comprises the three dimensions of space that define matter. But, without that fourth dimension, time, creation would be static. Nothing would ever change. Time has the unique property of making it possible for physical things to begin, develop, mature, regress and disappear. Even more than by space, our reality is defined by time. In God's reality (outside of our universe) there is no time. And that works, because (as was described above) He has no substance and He is a perfect unity who does not change. In order to put His new creations in a context that would allow them to accomplish something -- to change -- He provided the dimension of time.

Before something can come into being it must be conceptualized. The Kabbalistic system provides three *sephirot* that define the thought or conceptualization process. The *sephirot* of *chochma* (wisdom), *bina* (understanding) and *daat* (knowledge)[376] provide for the creation of an idea.

But it is only through the lowest (in order of their original emanation) *sephira* of *malchut* (kingship) that an idea can become a reality in our physical world. Malchut is the transitional *sephira*. While it "resides" in the spiritual, it occupies a "place" somewhere between the spiritual and physical worlds. Its role is to funnel the spiritual to the physical and the physical back into the parallel spiritual world.

---

[374] *A Brief History of Time*, p. 180-181

[375] Ramchal, Choker U'mekubal, Friedlander Edition, p. 7

[376] Daat is actually an operational substitute for the sephira of Keter.

Things exist in space, and space has three dimensions. By taking the parallelism between the spiritual and physical worlds to its next logical level we can conclude that our three spatial dimensions have parallel metaphysical dimensions. They are *chochma, bina* and d*aat*. These dimensions, three in each world, provide for the potential and existence of matter and place. Without them, neither "things" nor "ideas" could exist. And, without a fourth dimension in each realm -- time and its alter ego, *malchut* -- these spiritual ideas cannot become physical realities.

What about the other *sephirot* operating in the spiritual realm? Kabbalah describes them in vivid detail -- each as if it were a very human characteristic, "metaphysical" in every way. Is this where the idea of parallelism breaks down? How can there be a dimensional parallelism between 6 or 7 other spiritual *sephirot* and the four physical dimensions of our reality?

Hidden Dimensions

In constructing String Theory, a mathematical explanation that might lead to a unified theory of physics,[377] researchers substituted one

---

[377] Mathematicians continually come up with new ways to describe our world. Both real and imaginary numbers are used in their computations and simultaneously offer support for String Theory. Real numbers are basically one-dimensional. Complex numbers, which are a combination of both real and imaginary numbers, are two-dimensional. In the mid-19th century multiple dimensional systems of numbers were developed, among them octonions -- an 8 dimensional numbering system. ("The Strangest Numbers in String Theory," John C. Baez and John Huerta, Scientific American, May, 2011, p. 60) Supersymmetry was developed in the 1970s and 1980s. At its most fundamental, the universe exhibits symmetry between matter and the forces of nature. Every particle of matter has a corresponding particle carrying a force. And every force particle has a twin matter particle. So, the laws of physics would remain unchanged if we exchanged all of the matter and force particles. "Imagine viewing the universe in a strange mirror that, rather than interchanging left and right, traded every force particle for a matter particle, and vice versa. If supersymmetry is true, if it truly describes our universe, this mirror universe would act the same as ours."

Here is a more involved technical explanation from the same source: In an octonion universe, both matter and force particles "would be waves described by a single type of number, namely a number in division algebra, the only type of system that allows for addition, subtraction, multiplication, and division. In other words, in these dimensions the vectors and spinors coincide: they are each just real numbers, complex numbers, quaternions or octonions, respectively. Supersymmetry emerges naturally, dimensional "strings" for the tiniest of sub-atomic quantum particles. These strings are something like musical notes. They float in space and have some of the characteristics of actual strings, like tension and frequency. Physicists developed a concept called Supersymmetry that relates these particles/strings to one another, and in doing so identifies some as transmitting force while others comprise matter. All of this has been mathematically postulated so that a theory might be developed to account for gravity at sub-atomic levels (called quantum gravity). General relativity accounts for gravity in our realm, but at the sub-atomic level, the rules seem to be different.

Since there is no machine that can actually test it, String Theory has its share of doubters and a number of good reasons for doubt. Primary among them is the fact that in its original mathematical formulation String Theory required 26 dimensions. We ordinary, run-of-the-mill humans commonly recognize only the four mentioned previously, the dimensions of space and time. In the 1980's researchers managed to come up with seemingly workable theories that reduced the number of mathematically required dimensions from 26 to 10. Most recently, scientists have found a way to "roll up" the 22 or 6 extra dimensions as if they are there in nature, but are concealed from us. Something called M-Theory now seems to give increased validity to the 10 dimensional model.

Remember Hawking's explanation for why four dimensions works for us? Now he asks, "Why don't we notice all these extra dimensions, if they are really there? . . . The suggestion is that these other dimensions are curled up into a space of very small size, something like a million million million million millionth of an inch. This is so small that we

---

providing a unified description of matter and forces. Simple multiplication describes interactions, and all particles – no matter the type – use the same number system."

While the math allows for one, two, four or eight dimensions, when we include time in string theory and add it to space, we get supersymmetry in dimensions of three, four, six or ten. And string theorists have been saying for years that only 10-dimensional versions of the theory are self-consistent. Octonions provide the reason why the universe must have 10 dimensions. In 10 dimensions, matter and force particles are embodied in the same type of numbers – the octonions. Now membranes add another dimension, and researchers tell us that M-theory requires 11 dimensions (see the Kabbalistic relationship with the composition of the ancient ketoret/spices in Temple times).

The Elements of Jewish Exceptionalism

just don't notice it: we see only one time dimension and three space dimensions, in which space-time is fairly flat. It is like the surface of a straw. If you look at it closely, you see it is two-dimensional . . . but if you look at it from a distance, you don't see the thickness of the straw and it looks one-dimensional and highly curved. On bigger scales you just don't see the curvature or the extra dimensions."[378]

Are the other six dimensions of the spiritual world[379] paralleled in the physical world by being rolled up into the makeup of the human being? And putting on our philosopher's caps, we might ask, is it this hidden dimensional parallelism that makes it impossible for (reductionist) science to identify the true "rules of engagement" and predict human behavior in terms of mathematical models?

The four dimensions of our reality and their spiritual parallels are involved with the general administration of creation, either by God, or by the "forces of nature." But the lower *sephirot* and their parallels (the rolled up or hidden dimensions of our physical reality) seem to be microcosmic. The Kabbalistic tradition tells us that the spiritual dimensions are interactive. For a measure of a particular characteristic of existence to be "injected" into the physical world, the divine system works through the *sephira* or dimension of that characteristic.

For example, the Kabbalistic system tells us that the potential for kindness in creation has a spiritual source (an illumination) in the *sephira* of *chesed* (kindness). It is a channel between the immaterial and the created being. So, while Kabbalah does not talk about material reality, it does deal with the immaterial roots of that reality -- the *sephirot*. "Sephirot themselves are connected to each other and work through a blending of their strengths and qualities. The consequences of the sefirot and results of their actions are the totality of reality."[380]

While new discoveries and understandings in quantum physics are exciting, they tell us little about behavior. Instead we see that circumstances may predict a variety of possible outcomes, leaving us to

---

[378] *A Brief History of Time*, p. 179-180

[379] The other six sephirot are Din (judgement), Chesed (kindness, grace or love), Tiferet (beauty or mercy), Hod (splendor or majesty), Netzach (endurance or eternity) and Yesod (foundation or righteousness).

[380] *Song of the Soul*, p. 79

calculate the probability of each. In this sense, quantum mechanics introduced a new notion of randomness into science. Hidden dimensions in the Kabbalistic system track these developments and are a source for both scientific unpredictability and the notion that we will explore later -- free will.

How Science Views Creation: An Introduction

In 1900, the physicist Max Planck observed that a hot object emits heat radiation in little packets that he called quanta. Einstein applied this idea and terminology to particles of light (photons). But it became apparent that these "particles" also had wave-like properties, and as a result exhibited the behavior of both.

The field of sub-atomic physics offers ideas that are often at odds with tradition and common sense. Does it make sense that something can be both a particle and a wave? Or that a particle can be in two places at once? Can it be that sub-atomic systems become "entangled" so that the very measurement of one system has an effect on the other even if the two systems are light years apart?

Pioneers in this field have included Max Planck, Albert Einstein Niels Bohr, Werner Heisenberg, Erwin Schrödinger and many others. Heisenberg's Uncertainty Principle states that it is not possible for a quantum object to have a full and identifiable set of physical attributes at a given time. You simply cannot be sure of an exact position for and motion of a particular sub-atomic object. Heisenberg's principle defines the degree of this quantum ambiguity. But such "uncertainty" becomes less and less meaningful as objects and systems become bigger and bigger. So at our level, as humans, it seems completely irrelevant.

Because quantum processes are not necessarily traceable to a cause, they are called "irreducible." So, in a sense, they are truly random. At the quantum level, two identical sets of circumstances may produce entirely different outcomes. Heisenberg tried to define the odds of having one or another result.

An important feature of quantum physics is the role played by the observer of an activity or interaction. According to one of the primary schools of quantum thought, the Copenhagen school (Niels Bohr, 1927),

The Elements of Jewish Exceptionalism

the observation of an activity actually determines the outcome of the physical process that is being observed. In fact, it determines whether or not it takes place at all. If no one is watching, nothing happens! And while there are both waves and particles in quantum terms, one can act like the other depending upon the circumstances.

Erwin Schrödinger suggested that the "superposition" (a kind of overlapping) of the many possible wave forms or paths of an action would "collapse" (basically, consolidate on one of them) upon observation, yielding a single result.

As Heisenberg found, at the quantum level we deal with probabilities rather than certainties. In the quantum world, when we measure the state of one of a number of associated particles, constraints on one affect the others, even when there are large distances between them. This is called "non-locality."[381]

Sub-atomic physics tells us that the universe is quantized as Max Planck surmised over a century ago. It seems that energy, electric charge, mass and most likely time and space are not really continuous. They occur in the discrete units mentioned above called quanta. The space or intervals between these units is so small that this "choppiness" cannot be observed by us. Scientists understand that Newtonian physics cannot describe the behavior of particles at the atomic level. At that level, we are not describing particles at all, but instead probabilistic, statistical positions to which we are assigning names like "electron" or "proton." And, it is worth noting that a particle can go from one state to another without passing through any intermediate states. We are familiar with water changing from a liquid to a gas, but sub-atomic particles just disappear from one state and reappear in another. There is no transition.

Further impact on the relationship between the physical and what seems to be immaterial is evidenced by the existence of photons. These

---

[381] The experiments of Alain Aspect (1982) demonstrated that if two quantum systems interact and then move apart, their behavior is correlated in a way that cannot be explained in terms of signals traveling between them at or slower than the speed of light. This "non-locality" is either unmediated, instantaneous action at a distance or faster than light signaling. The theoretical groundwork for these experiments was done by Bohm and John Bell (CERN) and by the thought experiment of Einstein, Podolsky and Rosen in 1935.

are packets of energy of electromagnetic radiation (like light). The mathematical formulation that relates that energy to the frequency of the electromagnetic radiation includes something called "Planck's Constant." This fundamental constant of physics displays the incredibly tiny "scale" of quantum phenomena. A "Planck length" is derived by combining this constant with Einstein's speed of light and Newton's gravitational constant. Planck length is so small that it is about 10 to the 20th power smaller than the nucleus of an atom. It doesn't seem that Einstein's relativity (gravity) will work at the Planck length. Dividing the Planck length by the speed of light yields Planck time, about 10 to the -43 seconds. And because gravity is likely to work much differently before Planck time (at the big bang), the field of quantum cosmology was created to describe it.

A Theory of Everything? Determinism, Reductionism, Free Will and Quantum Cosmology

The universe is described by two theories of physics developed in the twentieth century: relativity on the "macro" scale, and quantum mechanics on the sub-atomic scale. Because these theories seem to operate in very different worlds, great effort has been expended in attempting to reconcile their differences. The search is on for a single, unified theory that ties them together.

If there really is a complete unified theory that governs everything, it presumably also supports the idea of determinism – that is the understanding that there can really be no such thing as truly free will. But this determinism must operate in a way that is difficult to calculate for an organism that is as complicated as a human being. We assume that humans have free will because we do not feel that we have the ability to determine behavior.[382]

According to the Kabbalah of the Ari, free will was introduced purposefully into creation with the "breaking of the vessels" described above. An overabundance of the light of the *kav* was temporarily contained in the lower seven of the *sephirot* (vessels) and it caused them to shatter. Even though they quickly repaired themselves to some degree,

---

[382] *A Brief History of Time*, p. 167

some of the original unity and order of creation was lost. It is from this spiritual event that imperfection -- the potential for evil -- was introduced into creation. And with it came the necessity for created beings to make choices – whether good or bad.

As described earlier, that mystical event also appears to have a theoretical, parallel in the physical, creation account. Inflationary theory [383] explains that after the big bang, particles expanded away from its epicenter very rapidly and at incredibly high temperatures. The four forces of nature that are dealt with by today's theoretical physicists were a single force at the moment of creation. But as expansion and cooling took place, there was a "phase transition," and the "symmetry" between the four forces (the strong and weak nuclear forces, the electromagnetic force and the force of gravity) was broken.

Back to String Theory: It's as if the bonds between the most elemental "strings" were now broken. The particles represented by strings are differentiated by their characteristics of vibration -- they play different songs. So the potential for the forces of nature to sing a song in unison was destroyed. That didn't preclude a nice harmony from being produced. But, according to Kabbalistic thought, it would take conscious effort on the part of created actors with free will (man) to ensure that this would happen.

In an expression of uniquely Jewish mystical theology, Kabbalah proclaims that the goal of creation and life is to re-establish perfection in the vessels – to re-establish unity among the strings, so to speak, and perhaps reunify the bedrock forces of nature so that a single song is sung by all creation. In doing so, evil and imperfection will be "returned" to good. That can only be done through the efforts of humankind. At that point, both divine and physical unity will be apparent.

---

[383] *New Scientist*, June 30, 2012, p. 32, 34, Amanda Gefter.
Inflation is a theory that makes the big bang work by ballooning the universe instantaneously in flatness and uniformity. Once inflation starts, however, it cannot stop. Bits of the inflating universe themselves begin inflating off into independent existences. This line of thought creates an infinite "multiverse" of universes, making cosmological predictions impossible. Inflation is a theory that is anchored in General Relativity because it is about space/time and gravity. It incorporates the uncertainty fluctuations of quantum physics as well.

The religious effort to reach this unification is mirrored in the scientific world:

Stephen Hawking wrote, "However, if we do discover a complete theory, it should in time be understandable in broad principle by everyone, not just a few scientists. Then we shall all, philosophers, scientists, and just ordinary people, be able to take part in the discussion of the question of why it is that we and the universe exist. If we find the answer to that, it would be the ultimate triumph of human reason -- for then we would know the mind of God."[384]

Steven Weinberg won a Nobel Prize for taking part in the development of a unified theory of the electromagnetic and weak nuclear forces. In his *Dreams of a Final Theory*, Dr. Weinberg covers a great deal of ground. "The standard model[385] of elementary particles now consists of Dr. Weinberg's electroweak theory and quantum chromodynamics, which describe the strong nuclear force. Gravity, the fourth known fundamental force, is subject to the very different principles of general relativity. Some physicists believe that it may be possible to formulate a theory unifying all four forces, a goal frequently labeled the 'holy grail of physics' or the 'theory of everything.'"[386]

Is a "theory of everything" possible? And if so, what would it tell us? Let's take a look at the intersection of physics and philosophy, the intellectual tools that are being brought to bear in this argument and how they reflect the inclusion of God in a solution.

Both philosophy and mathematics have been applied to the gap in understanding science and faith. The science of modern logic contributes

---

[384] *A Brief History of Time*, p. 191

[385] The guiding principle of the standard model is that its equations are symmetrical. This does not change even if the perspective changes or even when there is a shift in magnitude at different points in space and time. Matter is made of quarks and leptons influenced by three of the four known fundamental forces of nature: electromagnetism and the strong and weak interactions. Gravity is out for the moment. Quarks make up protons and neutrons which generate and feel all three forces. Leptons, like the electron, do not feel the strong force. Quarks have "color" and leptons do not. The strong force is carried by gluons, the other two forces by photons and three varieties of bosons.

[386] Professor Philip Johnson, University of California at Berkeley writing in the Wall Street Journal makes no claims to be an unbiased observer in this ideological battle and is known to be an ardent advocate of the Christian, religious right.

to this discussion. The mathematician, Kurt Godel, in what is known as his Incompleteness Theorem, proved that no self-contained system of logic (such as a theory of everything) is ultimately possible.[387] Of course the intellectually curious and cynical demand more.

Physicist, warfare analyst and philosopher James D. Sinclair analyzed the various scientific theories of physics and their implications regarding the existence and necessity of God. He reasoned that when one tries to reconcile the various theories of quantum creation and management with the very idea of God, both the Copenhagen[388] and Hidden Variables[389] theories (two influential theories of quantum cosmology) seem plausible and both require a necessary being – which is, after all, the definition of God. He looked at other theories and measured them against their compatibility with belief in the Divine. Superdeterminism (which completely disavows free will of any kind as discussed elsewhere) is by definition incompatible with divine faith. But Sinclair does not feel that it could possibly be true in any case. Other theories, among them Consistent Histories,[390] are not drawing away many adherents from the "established" big four (Copenhagen, Hidden Variables, Many Worlds, Superdeterminism) of quantum mechanics.[391]

---

[387] James Daniel Sinclair, "The Metaphysics of Quantum Mechanics," www.reasons.org. p. 10 of 13. Godel's theorems are important in the science of modern logic. They demonstrate the limits of provability in formal axiomatic theories.

[388] The Copenhagen interpretation of quantum mechanics was the first international attempt to understand the atomic world. Neils Bohr, Werner Heisenberg, Max Born and others were major contributors. It is primarily identified with quantum indeterminism today.

[389] Another theory of quantum mechanics attempts to explain away its probabilistic nature and things like "entanglement" by suggesting that there are variables that we just cannot recognize or measure.

[390] Consistent Histories, another explanation of quantum mechanics, tries to overcome the need for waveform collapse by assigning probabilities to the likelihood of various alternative histories coming to pass.

[391] John Bell is the author of "Speakable and Unspeakable in Quantum Mechanics: a one-world model that is holistic." A radioactive atom chooses to decay at a particular moment due to its interaction with every other particle in the universe. To us, who don't have access to this universal wave function, the decay seems random. The behavior is not random, it just seems so due to our lack of knowledge. Hidden Variables models must be non-local. Someone with full knowledge of the total wave function of the universe could non-locally impact any part of the universe meaningfully without

Each of these theories is colored by a philosophical predisposition. For example, in writing about his quest for a theory of everything, Steven Weinberg delivers a book about science, but with a "reductionist" philosophical message. And he offers this while simultaneously emphasizing the uselessness of philosophy in physics. In his view, all of the natural processes that characterize life are controlled by the genetic program in our DNA. And that genetic code is ultimately "reducible" to the laws of physics and chemistry that govern all matter. All of this and everything that has ever happened in the universe can be traced back to the initial conditions of the Big Bang. He concludes that our mental and spiritual life can be reduced to just chemical reactions in the brain.

Reductionist science and determinism leave no room for faith. Is the scientific search for a single unified theory handicapped by its own rules of inquiry?

SCIENCE AND BIBLICAL CREATION: A FAIRY TALE?

"In the beginning, God created the heavens and the earth."
"And God said let us make man in our image."
"The Lord God formed man from the dust of the earth. He blew into his nostrils the breadth of life . . ."

You have just given birth to your second child, little Emma, a younger sibling for your four year old son, Billy. Billy asks you to explain human procreation. "Mommy," he asks, "where did Emma come from?" Do you respond with an extensive anatomy lesson, or do you compose a story, heavy on meaning and results, but light on the details?

Now imagine that you are the creator of all existence. You have taken your time to bring into being a universe, for after all, time does not apply to you. In that universe you have allowed the development of a planet, ultimately occupied by living beings. Among those beings

---

detection. (Aspect experiments, See *Metaphysics of Quantum Mechanics*, p.5) So, if chance is real (Copenhagen interpretation) then God must exist as the cosmic observer. If determinism is real, then God exists as the Hidden Variable that stops the infinite regression of causes. There is also an approach that allows one to deconstruct reality: the Many Worlds and Superdeterminism models.

is a species – for whom you have designed creation -- that only after a long period of development is now sufficiently mature to embrace the knowledge of your presence. Until this point there has been only suspicion about your existence and power, both of which your creations have attributed to celestial bodies and other inanimate objects. You are now ready to formally introduce yourself to your creations, and in doing so establish your status as both their creator and the administrator of all that they see and experience.

In order to do so, you will need to provide a strong "back story," one that is both compelling and instructive. Doing so must take into account the primitive developmental state of your creations, emotionally, technologically, and sociologically. You must tell a story that retains the drama of creation and subsequent development, but is not burdened with incomprehensible detail. Since you are already an unknown entity to those creations, your story can logically be shrouded in mystery.

While too much detail may be counterproductive, your story must ring true. After all, you are the Master Being, and you must model behavior for your creations who you have molded in "your image." So your story is anything but a fabrication. Instead it is filled with objects, beings and actions that represent in a very real way a series of astonishing occurrences and story lines that explain the origins of all that has ever been known to be true.

That compelling, dramatic and instructive narrative must also be unforgettably fixed in the minds of your creations. So, you introduce it with special effects that sear it into the memories of witnesses that will repeat it, record it, discuss it and build their lives around it. The recording of this narrative is their instruction manual, and in it they will find all they need to know about the life you want them to conduct and the goals you want them to achieve. You make it clear that while stage, scenery and lighting are your doing; from this point forward the script is theirs to write.

In the Beginning

The Bible presents a series of seemingly inexplicable events which take place over a period of but a few days, culminating in the appearance of mankind. Both scientific discovery and the Kabbalistic religious tradition present this same series of events in what seem to be very different ways. Both utilize excruciating detail to explain or contradict events portrayed in the biblical text. The Kabbalistic narrative tries to honor that description with something tantamount to a "scientific" elucidation of that text. On the other hand, science does not pretend to honor the biblical text at all and instead presents a mathematical picture of events that it sees as contradictory to the biblical account. And yet, Kaballah created a vocabulary and hierarchy of terms as much as two millennia ago that correlate surprisingly well to twenty first century scientific and philosophical terminology and thought.[392]

These two schools of historical and derivative biblical analysis have deployed both physical and metaphysical terminology and metaphors to describe what we now understand to be the physics of creation. More often than not, the two worlds speak with a single voice. What did the Kabbalists know? And how, without modern science, could they possibly have painted so accurate a picture of both creation and theoretical physics?

An understanding of what happened "in the Beginning" is essential to the adherents of both science and religion. But, that understanding may be out of reach if we cannot relate to circumstances before the beginning. Without that frame of reference "the beginning" is described in an arbitrary fashion, subject to the predisposition of the analyst. Neither science nor religion is equipped to address this topic.

Instead, let's speculate. Imagine that before the beginning, there was infinite information -- no dimensions, either physical or temporal, no observer – only seemingly random bits of information, without order or relationships.[393] Had that information always existed? Of course, that question is meaningless. There was no "always." Time did not exist. Space-time did not begin until that information was manipulated or manipulated itself to produce a "space" in which a field of subatomic "matter" and/or "energy" could exist -- an implicate order in the

---

[392] See the regression analysis of Professor Hayim Shore in the language section above correlating the events of biblical creation with cosmology.

[393] Ed Fredkin's explanation of this follows in the section describing "digital physics."

nomenclature of physicist David Bohm (see below), or the *Tzimtzum* in the terminology of Kabbalah.

So, to paraphrase scripture, in the beginning God created the heavens (the ephemeral) and the earth (the substantive). And the earth (all that was now physical) was chaotic and formless.

With the "singularity," or the Big Bang, God -- the source or sum of infinite, pure information -- had established an ephemeral level of existence in which information was continually transformed into the most basic building blocks of matter. And from this substrate of the quantum field the elements were formed, and from the elements the physical structures that populate our universe.

Something from Nothing

The obtuse creation account begs an explanation that seems well beyond the ken of history and must surely depend on either fantasy or a formerly unavailable understanding of science. Blending our ever-growing scientific/quantum knowledge with our heretofore scientifically-independent faith has until recently been an insurmountable challenge.

Ancient cultures do not generally have a tradition of creation emerging from nothing at all, so it seems that it is religion, specifically Judaism, which introduced the idea. In fact the ancient Greeks had no word in their vocabulary for creation. Its translation of this term in Genesis is more properly "produced" or "made" than created.

The religious account of "the beginning" testifies to the creation of something from nothing. Is there a template for creation ex-nihilo? There is in mathematics! Using mathematical logic, the "axiom of existence" in set theory can be applied and leads one to assume the presence of an empty set.[394] Then, using the reasoning of Giuseppe Peano (early 20th century Italian mathematician and founder of mathematical logic and set theory) regarding the natural numbers, this empty set becomes the first natural number, zero. A subsequent set whose sole member is the original empty set gives us the next natural number, 1, and so on.

---

[394] Mathematicians, Ernst Zermelo and Abraham Fraenkel's axiomatic set theory offers a theory of sets free of paradoxes. The theory uses first order logic to arrive at the conclusion that if there are sets at all, then the axiom of subsets tells us that there must be an empty set: There is a set such that no set is a member of it.

So the idea of an empty set allows for the creation of the infinite set of natural numbers. The infinite series of natural numbers derived in this fashion is an example of creation "ex nihilo" – natural numbers that are born of an empty set.[395] But this is a bit too academic for most.

In their book, *The Grand Design*, Stephen Hawking and Leonard Mlodinow have a very different position on the sudden appearance of existence. "It is possible to answer these questions (about creation) purely within the realm of science, and without invoking any divine beings." The universe arose "from nothing" due to the force of gravity. The laws of nature are an accident of the particular slice of universe that we happen to inhabit. They base their conclusions on the development of String Theory and M-Theory.[396]

Something from nothing? The Big Bang theory proposes that the universe evolved from a single, pre-existing, infinitely dense point at which space and time did not exist. Then, after an inexplicable explosion, space, time and matter came into being. At the first moment, the four fundamental forces of nature (gravity, electromagnetism, and the strong and weak nuclear forces) were unified. This super-hot and dense point continued expanding rapidly so that one millisecond after the big bang the universe was thirty million times hotter than the surface of the sun today and fifty million times denser than lead. During this "inflationary" period, the universe doubled in size one hundred times in less than one millisecond. At the age of only one one-hundredth of a second, the link between the four fundamental forces was broken. Light remained trapped in the ball of energy and matter could not yet survive.

One second after the big bang, the universe had grown to twenty light years in size. The temperature had cooled to ten billion degrees

---

[395] B'Or Ha'Torah 17, p. 118, "Creation ex Nihilo, Number Theory, Quantum Vacuum, and the Big Bang," Alexander Poltorak.

[396] String Theory offers the possibility that our 4 dimensional reality is situated on a "brane," a lower-dimensional object floating in a higher-dimensional space. Ours is not necessarily the only brane because two branes floating a microscopic distance from one another can form the boundaries to a 5-dimensional space in between, like a sandwich. The branes touching resulted in an explosive exchange of energy, a singularity occurring as the 5th dimension briefly disappeared. According to this theory, we interpreted this event as the big bang – not the beginning of our space/time, but a re-energizing of an eternal cosmos. This is called the Cyclical Model by Paul Steinhardt of Princeton.

and after three minutes to one billion degrees. At this point matter could begin to form. For the next 300,000 years the universe continued to cool and expand, the temperature reaching 10,000 degrees Kelvin. Helium atoms appeared and light began to become visible. The universe has been expanding ever since.[397]

BIBLICAL AND SCIENTIFIC AGREEMENT ON THE AGE OF THE UNIVERSE

Science has routinely denigrated the Bible's story of creation and its chronology. Yet, with the support of Einstein and relativity, the Big Bang has become a popularly accepted fundamental of not-so-theoretical physics. The Big Bang and the biblical account of "something from nothing" are, at least conceptually, describing one and the same event – an event described by the biblical commentators, Nachmanides in the 13th century and Rabbi Isaac Luria[398] in the 16th century, as the emanation of creation from a single point of energy. But even this agreement does nothing to scientifically reconcile the billions of years since the Big Bang with the few thousand years that are derived from biblical chronology. How can any rational, modern man believe that the world was created in six, 24-hour days just a few millennia ago?

At the most basic level of scientific and rational thinking, our quasi-scientific and evidentiary knowledge of the universe is more than enough to make purely biblical statements about creation seem ludicrous. Further examination may yet contradict that conclusion.

---

[397] B'Or Ha'Torah, Vol. 13, p. 23-24, Alexander Poltorak, "On the Age of the Universe"

[398] Rabbi Isaac Luria (Ari, sixteenth century) is quoted in *Eitz Chaim* as saying, "When it arose in God's Will to create worlds and emanate the emanated . . . He contracted Himself in the point at the center, in the very center of His light." He then caused an "explosion" which initiated the universe. Nachmanides (13th century) states in his commentary on Genesis, "The Holy one, blessed be He, created all things from absolute non-existence. Now we have no expression in the sacred language (Hebrew) for bringing forth something from nothing other than the word *bara*. Everything that exists under the sun or above was not made from non-existence at the outset. Instead He brought forth from total and absolute nothing a very thin substance devoid of corporeality but having a power of potency, fit to assume form and to proceed from potentiality into reality."

The reconciliation of billions of years on the one hand with a few biblical days on the other has been addressed in many ways.[399] The Midrash Rabbah on Leviticus 29:1 explains that the six days of biblical creation are a calendar unto themselves. Time after the creation of Adam on the sixth day differs from what came before, in which case, billions of years may have passed before the advent of man. Nachmanides (Rabbi Moshe ben Nachman, 13th century Spain) comments on the first chapter of Genesis, remarking that the first day of creation is not called the "first" day in the Hebrew text, but rather "one day." So the Bible is telling us of the creation of the concept of time. He comments there that the miniscule dot from which all of creation emerged was not what we think of as matter, but was so "thin" that it really had no substance at all. Only when matter formed from this substance did time begin. As we will later see, this is eerily descriptive of the now accepted scientific description of creation and it expresses on a quantum level, the substance of creation.

Still other explanations are offered for apparent contradictions between the biblical presentation and the scientific explanation. For example, each of the six days of creation might have been longer than a 24 hour day. According to Psalm 90, a thousand years are like a day to God. But that is still a far cry from the billions of years that scientists tell us are to be taken into account.

Each of the six days of creation was in fact 24 hours long, says the Babylonian Talmud (tractate Chagiga, page 12a). But those six days also contained "all the ages of the world." And that explanation may presage Einstein's theory of relativity which described a universe in which the rate at which time passes varies under different circumstances. Two clocks traveling across a continent, one by car and the other by airplane, would measure time differently. Time would actually progress more slowly on the plane than in the car on the ground. As a result, if you wanted to hold back the aging process, you could try flying continually at supersonic speed from west to east. "The theory tells us that as velocity or gravity in one place increases relative to a second place, the

---

[399] For one example, see Chaim Shore's calculations in the Hebrew language section above.

The Elements of Jewish Exceptionalism

flow of time at the first location slows relative to the flow of time at the locations where the velocity or gravity remained low."[400]

Time Dilation, as this phenomenon is known, allows time to slow as an object races through space. An imaginary space ship travelling at nearly the speed of light would see one of its days equal more than a month back on Earth. And because this relationship applies to both velocity and mass, the passage of minutes would be the equivalent of earth years on a planet whose mass is many times that of earth. Extrapolate that idea to the mass and the radius of the entire universe and one can start to grasp the idea that six days of creation might have been quite a bit longer than the days that we experience on earth today. Physicist and writer Gerald Schroeder uses this approach to calculate the time dilation caused by the difference in gravity between two locations. He concludes that the passage of one minute at the theoretical "edge" of the universe would be experienced as the passage of a million-million minutes on earth. "In terms of days and years and millennia, this slowing down of time by a factor of a million million reduces 15 billion years to . . . six days!!!"[401]

An analysis of temperature change offers another explanation. After the Big Bang, temperatures were so incredibly high that all matter was in the form of photons traveling at the speed of light. At that speed, time cannot "grab hold." But as the universe began to cool, sub atomic "quarks" became confined to the protons and neutrons that now comprise all matter. The clock of both the material universe and the Bible begins at this quark confinement. As the universe has continued to expand, its temperature has dropped from the threshold temperature that allowed the formation of protons and neutrons (a nearly unimaginably large number) to only 2.73 degrees Kelvin. The difference between these two temperatures allows for the derivation of time in both the biblical and scientific analyses. According to this argument (which echoes the Talmudic statement from tractate Chagiga), six 24-hour days at what

---

[400] *Genesis and the Big Bang:* Dr. Schroeder develops this argument even more convincingly in his *Science of God*. There he actually correlates the six days of creation with the billions of years since creation and earth's history as recorded in fossil records.

[401] Ibid.

is currently the outer limit of the universe would be about 15 billion of our years from our frame of reference on earth.

As an aside, many physicists now believe that the concept of time may be the key to unifying relativity and quantum physics. Researchers are attempting to offer a "timeless" theory to address this question. That is, time may not exist at all at a fundamental level, but may arise only at higher levels. ". . . just as a table feels solid even though it is a swarm of particles composed mostly of empty space. Solidity is a collective, or emergent, property of the particles. Time, too, could be an emergent property of whatever the basic ingredients of the world are."[402]

Biblical exegetes have calculated the age of the universe in varying ways but in every case come up with similar conclusions. Some of these approaches are based on the Sabbatical cycles mandated in the Bible both for the replenishment of the land and the conclusion of business and personal contractual arrangements. After seven of these seven year cycles, the fiftieth year is declared the Jubilee year.

According to the Gemara, "the world will exist for six thousand years and in the year 7000 it will be destroyed."[403] This describes a grand Sabbatical cycle of 7000 years. The Mishnaic sage, Rabbi Nechunia ben haKanah declared in his book, *Sefer haTemunah*, that the world will exist for one Jubilee cycle, or 49,000 years in total. If we are currently in the last Sabbatical cycle of that Jubilee, and the first modern man was created at the end of the first Sabbatical cycle, then humankind *as we know it* would be 42,000 years old.[404] This is consistent with anthropological research.

There is also a tradition that "God built worlds and destroyed them" before he created the universe as we know it.[405] The *Zohar* notes (as do many other biblical commentaries, and of course, scientists) that other species of man already existed before Adam (modern man) was

---

[402] Scientific American, June 2010, p. 60, "Is Time an Illusion?" Craig Callender

[403] Talmud Bavli, Sanhedrin, 97a

[404] *Evolution: the First Four Billion Years*, McHenry, H.M., "Human Evolution." Man first evidenced both anatomical and behaviorial modernity about 50,000 years ago.

[405] Bereishit Rabbah, 3:7

created. [406,407] And, the midrash teaches that there were "orders of time" before the creation of this universe. These traditions are consistent with the biological development of modern man's precursors as well as the physical development of the universe.

Both the Gemara and a number of midrashim state that there were 974 generations before Adam.[408] Since a phrase from psalm 105 is taken to indicate that the Mosaic law was to be given to mankind after 1000 generations, and Moses is the 26th generation from Adam, 974 generations must have come before. "We see from here that there was a system of time beforehand (that is, before Adam)."[409] The Kabbalist, Rabbi Yitzchak ben Shmuel of Akko (13th century) noted that since these cycles existed before the creation of Adam, they must be measured in divine years, not human years. In the words of one of the Rishonim (medieval biblical exegetes), "One of God's days is a thousand years as it says, 'For a thousand years are in your eyes like a yesterday.' (Psalm 90:4) Since one of our years is 365 ¼ days, a year on High is 365,250 of our years."[410] If we multiply this divine year by the 42,000 years that have passed in the Jubilee, we find creation to have taken place about 15 billion years ago. This too is consistent with the mathematics of the Big Bang.

One might then ask if the night and day referred to in the biblical account of creation refer to solar periods as they do today. Since the sun was not created until the fourth day, this definition may not apply. It may be that as the Midrash states, "until (the sixth day when man was

---

[406] Zohar, Vayikra 10a

[407] Scientific American, May, 2013, p. 66: As recently as 40,000 years ago Homo sapiens lived alongside several kindred forms, including the Neandertals and tiny Homo Floresiensis. Genetic studies in the 1980s led to a popular theory that anatomically modern humans arose in Africa and spread out across the rest of the world, replacing the existing groups. They seemed to have eliminated their competitors without interbreeding with them (the African Replacement Model). But recent DNA studies show that people today carry DNA from Neandertals and other archaic humans.

[408] Talmud Bavli, Chagiga, 13b

[409] Rabbeinu Bachya, Commentary on Genesis, 1:3

[410] Isaac of Akko, Otzar HaChaim, pp. 86b-87b

Unholy Alliance—The Scientific & Religious Conspiracy

created) the counting is done relative to the universe. From this point forward a different system of counting began."[411]

Unfortunately the order of creation also presents a problem. It is clear that the stars must have preceded the earth and its water, that vegetation post-dated the luminaries, and that land creatures preceded birds. Even the biblical commentator Rashi points out that "the text is not intended to point out the order of the various acts of creation."[412]

It is likely, then, that the actual details of creation are both scientifically and logically beyond human comprehension. Even the incredible sub-atomic knowledge that we now possess may be insufficient to explain the intent and source of creation, though it may teach us about its composition.

The Rambam (Maimonides, 12th century Jewish exegete) may have believed that there were not separate creative acts on six days but that everything was created on one day, in a single instant. "In the work of creation it mentions six days to indicate the different levels of created being according to their natural levels of importance, not that there were actual days, and not that there was an ordered sequence to what was created . . . He (Maimonides) tried to conceal this view with ingenuity, as can be seen in his words there."[413]

Scripture itemizes steps taken in the creation process that reflect scientific reality. An example is the conventional wisdom that says billions of years passed in which random reactions changed water and elements into organisms.

In the 1970s E. Barghoon and Stanley Tyler discovered micro-fossils of bacteria and algae in chert rocks (a kind of silicon dioxide) 3.6 billion years old -- that is just after the time liquid water appeared on earth. They concluded that life appeared immediately after water, exactly as the biblical account would have it.

On day 3 the word *creation* is not mentioned in the Bible. The universe was equipped with the necessary components from inception so it was not created, but formed. Day four has no mention of the development of life. The fossil record also has this gap. Life remains single

---

[411] Midrash Rabbah, Breishit, 9:14

[412] Rashi's commentary on Genesis 1:1

[413] Abarbanel, Genesis commentary

celled for another 3 billion years until the Cambrian explosion, at which time every basic animal body plan (the 34 animal phyla) that we have today appears in the fossil record.

Even if the six days of creation were actually six, twenty-four hour days as Nachmanides (13th century Spain) insisted, we may want to fall back on the Kabbalistic argument advanced above: "But in its real essence . . . the text has quite a different connotation. It refers to the six sefiros, which are modes of revelation of the Divine conduct of the world. Only for our benefit does Scripture present them to us in the form of six days. As for the relevance of the six days in their allusion to the six modes of revelation – this is something sublime and concealed from us, as Ramban (Nachmanides) says."[414]

THE IDEA OF CREATION BECOMES CREATION

There is also a classical, religious basis for the belief that in His acts of creation, God first laid out a spiritual plan and only then followed with its physical implementation.[415]

Physicist, rabbi and author, Aryeh Kaplan allows that "It would appear then, that the seven days of creation described in the Torah actually occurred in thought rather than in deed . . . thus it may be that the seven days of creation took place over 15 billion years ago, before the Big Bang. This represented the creation of the spiritual infrastructure of the universe, which the Talmud refers to as 'creation in thought.' The universe then developed according to God's plan, guided by the spiritual infrastructure He had created."[416]

Following this reasoning, on the sixth day of creation, God infused human beings with a spiritual sensitivity and intellectual capability that made them in His "image." This folds very nicely into the "superior soul" theory: even in the presence of other humans, Adam was distinguished in that God made him the first of his species to possess the power of prophecy, or the ability to receive Divine revelations.[417] This

---

[414] *Michtav Me-Eliyahu*, Vol. II, p. 151

[415] Talmud Bavli, Berachot 61a, Eruvin 18a

[416] *Immortality, Ressurrection and the Age of the Universe: A Kabbalistic View*, p. 11

[417] As suggested by Nachmanides in his commentary on Genesis 1:26-27

conclusion is supported by the use of the Hebrew terminology "ruach Elokim" (Genesis 1:2) which in dozens of other appearances in the Bible refers exclusively to prophecy.

Yitzchok Block introduces an approach that each of us has speculated about at one time or another: "Imagine that by dreaming one could create entities that have an existence in their own right... Let us imagine for the moment that our dreams do have genuine creative power that can bring into existence entities that have a reality of their own that could include real people and things, and that these entities cease to exist when we wake up.[418] As far as we know, such dream worlds might indeed be created by our dreams and destroyed by our awakenings in a manner such that we have no awareness of them[419]... (the creations in your dreams) are unaware that their existence depends on your dreaming them and that they will cease to exist when you wake up from your sleep. How could one possibly explain to the (creations) their true situation in a way that they could understand? Wouldn't (the creations) say that this is utterly preposterous and there is absolutely no empirical evidence for this wild idea?"[420] The dream is God's creation of the universe, and while it has no real existence it is very real to the creations whose existence continues to depend on an immaterial creative act taking place on a higher plane.

If this elaborate template was dreamed up and implemented by the Creator before the actual physical act of creation, then "Given the entire functioning integrated blueprint of a universe containing moral beings, a big bang could then be designed and programmed to teleologically produce them... (Before man's emergence everything is preprogrammed)

---

[418] "Idealism... suggests the thesis that bodies or material or physical objects are merely ideas or objects of thought or perception." Scientific American, Jan/Feb, 2014, p. 35, "Idealism Vindicated," Robert M. Adams

[419] Hofstadter believes that identity comes from a physical substrate, not limited to the brain, which is composed of neurons, which are in turn composed of molecules, and so on all the way down to quantum particles – which for Hofstadter, is where the true causality of the physical universe ultimately resides – without any recourse to the metaphysical or supernatural dualism. We each "hallucinate" a self. He sees I as an hallucination perceived by an hallucination. We believe the story we tell ourselves – we believe in a myth. (*I am a Strange Loop*, Douglas Hofstadter, p. 295)

[420] Yitzchok Block, "Creation: The Argument from Finitude," B'Or Ha'Torah,, no. 11, p. 15

an acting out of the mechanistic laws of nature with some quantum randomness thrown in . . . in the teleological sense, creation is completed not with the emergence of the Big Bang but rather fifteen billion years later when the first intelligent moral being emerges and decides to accept the burden of moral responsibility for its actions."[421]

But why would the Creator dream up or "design so cruel and wasteful an evolutionary system leading to the ultimate goal?" The answer to that question may be that ". . . when the evolutionary process is seen as the 'computational device,'. . . it can be seen in all its elegance. Evolution by random mutation in this sense is a self-improving program. It is a very simple yet efficient algorithm, used to run the 'computer simulation' leading up to the evolution of ever more complex creatures. Similarly, the Big Bang theory is a beautifully simple algorithm for generating a blueprint of an extremely complex universe. Given the design of the intended moral being, the big bang generates a complex universe . . . a singularity . . . operating according to one unified universal law in a four (or perhaps higher) dimensional space-time, and the rest takes care of itself." Avi Rabinowitz continues, "By mentally extrapolating this algorithm, God obtains very simply a complete description of a totally self-consistent complex universe and uses this description to create an actual universe at the moral stage without any (destructive, violent) physical evolution."[422]

SCIENTIFIC CONCLUSIONS ABOUT CREATION

The earliest philosophers, Aristotle among them, asked a very simple question: need there have been a beginning at all?

While twentieth century physics suggests an answer in the affirmative, modern cosmologists and physicists have been working overtime to suggest not. Reviewing the terminology and concepts of myriad competing theories yields a sense of the kind of mathematical and conceptual thinking that is being brought to bear in an attempt to explain

---

[421] Avi Rabinowitz, PhD, "And God Said, Let There Have Been a Big Bang, B'Or Ha'Torah, No.13, p.10

[422] Ibid, p.12

the unimaginable. These "models" of quantum cosmological physics were mentioned above and are reprised here.

The Many Worlds[423] hypothesis is employed by Stephen Hawking to obviate the necessity of a "beginning" and account for the apparent fine-tuning of creation (the Anthropic principles, the Teleological Argument for God). At the same time he offers an appeal based on the intrinsic randomness of an uncaused beginning (quantum fluctuation). This – chance -- he believes is how the whole thing got started. Yet, unlike the Greek notion of pre-existing matter, it does necessarily demand a "beginning" of some kind.

In quantum mechanics, the Copenhagen[424] and Many-Worlds models directly contradict one another. The former sees every observation resulting in an event while the latter is observation-independent. And even the Many Worlds model is subject to dispute among its supporters. Hugh Everett, the first to propose the Many Worlds theory in 1957, could not choose between Copenhagen and Many Worlds and was convinced that reality began as one universe which branched out as necessary every time a quantum event occurred. This approach eliminated the measurement problem (the idea that measurement or observation of something collapses its wave function and creates the reality we observe)[425] of the Copenhagen school because a collapse of the wave function never happens. All well and good, but this model does not solve the problem of a beginning.

---

[423] Where a universal wave function obeys the same deterministic, reversible laws so there is no wave function collapse associated with an observer. Instead, the universe repeatedly splits into mutually unobservable alternate universes within a multiverse.

[424] The Copenhagen interpretation of quantum theory was agreed upon at the Solvay Conference of 1927. It states that the wave function represented our knowledge of a system and the particle that is the result of the collapse of the wave function represents an even more precise knowledge of the system due to measurement. The knowledge spoken about here is mathematical theory, representing the knowledge of the human observer, not real world physical knowledge. The theory represents the work of Bohr and Heisenberg.

[425] In addition to Many Worlds, another possible solution to the measurement problem states that the macro-world does not suffer from quantum ambiguity because it can store information and is subject to the arrow of time. The micro world cannot store information and is time-reversible (H.R. Pagels, *The Cosmic Code: Quantum Physics as the Language of Nature*, New York, Bantam, 1983).

Bryce Dewitt, a pioneer in quantum gravity at the University of Texas, strays even further from the cosmological narrative by saying that many worlds have always existed. There is no beginning because time itself doesn't really exist. Reality is like a series of still photos that last eternally. Time only exists when the photos are collected and arranged in a linear sequence. The laws of physics are the glue that holds the sequence together and in order.[426] DeWitt's math seems to show that quantum mechanics and gravity (general relativity) are reconciled when time drops out of the calculations completely. But not so fast! If we believe that the past is an illusion, then how can there be cause and effect? With no "pre-occurring" cause, why do events occur at all?

In his argument, Stephen Hawking eliminates the need for a beginning "in time" by proposing that reality is in a "closed time loop." For times beyond the miniscule Planck time,[427] the universe expands out of a Big Bang until gravity halts that expansion, and then the universe contracts in to a Big Crunch. For times that are so small that they are near the Planck time, time begins to act as a true spatial dimension.

The Hawking model[428] is cyclical in time so a big bang must become a big crunch. And the condition of the singularity at the big bang and big

---

[426] *Fabric of Reality*

[427] In 1899, German physicist Max Planck proposed a group of natural units which are defined by properties of fundamental physical theories rather than by experimental parameters. These are universal constants that include gravity (G- general relativity and Newtonian gravity), the Planck constant (h, quantum mechanics), the speed of light in a vacuum (c, electromagnetism and special relativity), the Coulomb constant (k, electrostatics), and the Boltzmann constant (kB, thermodynamics). Planck's constant is a fundamental feature of quantum physics and relates the frequency of radiation to its quantum energy. It has the dimensions of action (energy x time).

[428] What if we look at the development of the universe as a wave function? Looking backward from the present, there are many possible paths that could have been taken. Then the starting point is not the beginning of time in a singularity, but a timeless point where the superposition of all histories explodes into existence from nothing with the existing laws of physics. This is called the No-Boundary Proposal by Stephen Hawking and James Hartle. Here the understanding is that there is a single universe with multiple histories. The resulting wave function removes the measurement (observer) problem and spells out the unique set of probabilities for anything that might be observed. The results that we see today become the input to the theory (flatness of space/time, homogeneity of the CMB). The probability of inflation having occurred using this model is very high (Physical Review Letters, vol. 100, p. 201301, Hartle, Hawking, Hertog, 2008).

crunch must be identical. This means one of two things must be true, and both are unacceptable to Hawking: either the entropy (see earlier discussion) at the singularity is zero, in which case Hawking must find an entropy reversing process in nature, or the entropy at the singularity is some big number, in which case Hawking must explain how an unintelligent process somehow hit the bulls-eye in producing a life-giving universe (the initial condition problem).

Theoretical physicist and cosmologist Alan Guth has shown that the entropy reversing process is impossible. He explains it in this unappealingly technical way: "Or, Hawking must admit he has a theory with a quantum singularity (an infinite collection of 4-spaces each with zero volume). This would permit his outgoing and incoming world lines to meet at the singularity without being continuous in entropy. In this case, Hawking has admitted the existence of a boundary to his universe which is uncaused and has created things with lower ontology (God by another name)."[429] The independence that Hawking is out to prove seems to have escaped him.

Got that? The point of this discussion should be obvious: despite Hawking's best intentions, he hasn't come up with a convincing way to dismiss the need for a beginning.

We've reached the nexus of science and philosophy. For now, it helps to understand that the biggest problem with the Many Worlds model is the rationality problem, expressed well by Canadian philosopher, John Leslie:[430]

If all possible things routinely happen within the multiverse, then why do we live in a rational universe? Events with low probability, like the origin of life, can only be explained by appealing to the Many Worlds rationale (in an infinite number of worlds, life must be possible on at least one of them!). But this probability is less than that of the appearance of a perpetual motion machine (a probability near zero because it is precluded by the laws of thermodynamics).

Scientific atheists require the use of Many Worlds to counter the problem of the observer (basic to the Copenhagen interpretation of

---

[429] Sinclair, p. 8

[430] *Universe*, John Leslie; also see varying other critiques, among them those of Lubos Motl of the Czech Republic

quantum physics), the problem of the beginning (as you see above with Hawking) and the problem of fine-tuning (otherwise known as the case for design -- one of the philosophical/logical proofs of Divine creation). But if Many Worlds produces an infinite number of universes, this produces a necessary rather than contingent reality. A necessary reality implies a creator. Philosopher (and converted atheist) Anthony Flew said that "from necessary things, only necessary things come." If he is right, then the universe was arguably something necessary that came from a necessary being.

But there is more! Roger Penrose of Oxford published a paper echoing Hawking's cyclical big crunch hypothesis, his focusing on cosmic microwave background radiation (CMB) and pointing to its irregularities. He believes that it is more uniform than it should be and that this uniformity betrays traces of black holes from the pre-Big Bang version of reality – the something that came before there was something. He believes that the uniformity shortly after the Big Bang came not from inflation as most of today's theorists advance, but from the tail end of reality's previous incarnation. The universe came into existence through a continuous cycle of iterations. Each starts with zero size and great uniformity. It becomes less uniform as it evolves and objects form within it. But after enough time all the matter that is around gets sucked into a black hole. Those black holes eventually evaporate in a burst of radiation (called Hawking radiation). This process increases uniformity, eventually bringing it back to the level at which the universe began. Penrose believes that when the universe becomes very old and rarified, all particles lose their mass. Massless particles would travel at light speed, so the particle would experience neither space nor time. There would be an infinitely small universe, one that would then undergo a Big Bang.[431] That's the way it began, ended and began again.

It should be obvious that we are a long way from unanimity on these topics. The technicalities of the arguments are bewildering. However, each potential explanation offers a new set of tools to apply to the concept of creation and those tools bring us closer to the very explanations (some of which have already been presented) that are offered by the Kabbalistic tradition.

---

[431] *The Economist,* December 4, 2010, p. 101

INFORMATION AND THE MYSTICAL TRADITION

At the risk of offending generations of traditional proponents of mysticism, I offer that Kabbalistic thought and interpretation is neither magic, nor mystical. It is mathematical. Hebrew letters have quantitative value and the words that they form have meaning that is directly related to that numerical value (see the earlier Hebrew language section). The science of extracting data from the Hebrew language was once understood to be the essence of Kabbalah. The 18th century Italian Kabbalist Moshe Chaim Luzzatto (Ramchal) divined the exact dimensions of the third Temple, as described in the prophecy of Ezekiel, from the language of that prophet. His *Mishkeney Elyon* is a text entirely dedicated to this topic.

History demonstrates that from time immemorial philosophers have also associated mathematics, the substance of both physics and information science, with the divine.

For example, "(Examine) ... the ideas of the Neo-Platonist Proclus, a follower of Plotinus and one of the last representatives of philosophy in Antiquity. While Nicomachus used numbers to connect the world and the divine, Proclus used geometry in order to lead the pupils in his school in Athens to the divine. For the Neo-Platonists transcendence was characterized by an immediate and total unity of the thinking subject and its object. In Proclus' view the soul can attain such a self-discovery in geometry. In its geometrical projections the soul sees an image of itself and can thus attain a knowledge of truths concerning transcendent first principles, the gods."[432]

Today, we draw similar conclusions, not from philosophy, but from the natural sciences: "Physicists have, over the past couple of decades, uncovered profound insights into how the universe stores information – even going so far as to suggest that information, not matter and energy, constitutes the most basic unit of existence. Information rides on tiny bits; from these bits comes the cosmos." [433]

---

[432] *Mathematics and the Divine: A Historical Study*, p. 18

[433] "Is Space Digital," Michael Moyer, *Scientific American*, February, 2012, p. 31, 32. "Hogan's interferometer will search for a backdrop that is much like the ether – an invisible (and possibly imaginary) substrate that permeates the universe. By using two Michelson interferometers stacked on top of each other, he intends to probe the

The Elements of Jewish Exceptionalism

Computer researcher and pioneer of Digital Physics, Ed Fredkin explains, "... it may be that information processing, instead of being the sole province of us humans and our machines, may be a part of almost everything else in physics. Life itself, is clearly mediated by digital information -- the genetic code. Digital Mechanics assumes that physics is also a process based on informational processes... We should not be afraid to consider intellectual activity as the driving force behind the creation of the Universe."[434]

And this is entirely consistent with the Kabbalistic model. It may be that in the course of creation as well as on an ongoing basis, God introduced His information in binary form, i.e., as cellular automata (as Fredkin would propose).[435] These expressions of data generate what physicists call the quantum vacuum – a kind of field of infinite energy.[436] The automata (acting as software) produce action in the quantum vacuum out of which the physical world and even consciousness emerge. This divine introduction (referred to in the literature of Kabbalah as the

---

smallest scales in the universe, the distance at which both quantum mechanics and relativity break down – the region where information lives as bits." P. 34

"Yet the Planck length is much more than the space where quantum mechanics and relativity fall apart. In the past few decades an argument over the nature of black holes has revealed a wholly new understanding of the Planck scale. Our best theories may break down there, but in their place something else emerges. The essence of the universe is information, so this line of thinking goes, and the fundamental bits of information that give rise to the universe live on the Planck scale." p.35

[434] "A New Cosmogony," Edward Fredkin, www.leptonica.com/cachedpages/fredkin-cosmogony.html

[435] A cellular automaton is a group of cells on a grid of a specific shape that evolves over time in steps according to rules that are based on the "states" of the neighboring cells. The rules can continue to be applied for as many time steps as desired. They can progress along a shape of one or more dimensions. The states of the cells are represented by colors. The cellular automaton "lives" on a grid with cells of a certain "color." The "neighborhood" in which the cells live influences them. At each time step, neighbors will affect the adjacent cells. Common neighborhoods in two dimensions are named for developmental mathematicians, the Moore Neighborhood for a square grid and the von Neumann Neighborhood for one that is diamond-shaped.

[436] A quantum vacuum is not simply an empty space at the very foundation of being. Instead there are electromagnetic waves and particles that pop into and out of existence from and within it. Even though the quantum field has no photons, it has infinite energy due to the possible frequencies of electric and magnetic fields that may be present in a sort of energy "bath."

Or Ganuz or Supernal light) is pure information. This idea was known in religious intellectual circles by Maimonides – not thought to have Kabbalistic leanings -- who called this intellect or the divine overflow. The information/intellect which becomes the substrate of all else is not subject to the restrictions of time. However, the quantum particles that arise from it are. In this thought proposal, information is the original, supernatural emanation, manipulated by the software within the Kabbalistic *sephirotic* structure.

A "way out there" explanation for existence? Perhaps, but put aside the technical aspects of this approach for the moment and consider just how much sense it makes. Whatever is going on at the sub-atomic level is both immaterial and incomprehensible to us. So doesn't it make perfect sense that the essence of the universe is something even less material than that? There must be a substrate to the universe at the very smallest scale, where quantum mechanics and relativity both break down and all that exists is information.[437] Physicist Leonard Susskind explained in a lecture at NYU that information actually means distinctions between things and that a basic principle of physics is that distinctions never disappear, despite the fact that they may become scrambled. So, there is good reason to believe that all existence is not just based on information, but is information. And it is through its manipulation that reality emerges.

The overlap of science and faith is then unmistakable. The biblical God defines Himself in the language of informatics, if you will. "I am that I am,"[438] is God's response to Moses when asked for His name. God describes His own state of existence as the product of a yes/no, 1/0, Am/am not binary decision. In the vernacular of computer science, existence is one bit, that is, one piece of information.

"Take the two hydrogen atoms and one oxygen atom of the water molecule. When they merge, they calculate the optimal angle and distance at which to attach. All possible paths are analyzed and the optimal angle of 104.45 degrees is typically chosen. This is like a mathematical simulation of the physical reality, and yet it is reality," said John Archibald Wheeler. "The universe is a mathematical simulation, and

---

[437] *Is Space Digital*, p. 35

[438] Exodus, 3:14

that simulation is reality. And what is the computer that conducts the simulations? The universe itself. The purest state of existence is the choice of here/not here or am/not am."[439] "Every it – every particle, every field of force, even the space-time continuum itself – derives its function, its meaning, its very existence entirely from binary choices, bits. What we call reality arises in the last analysis from the posing of yes/no questions."[440]

The manipulation of information is computation. To initiate that process, both information and energy are required. From these inputs order emerges, along with structure and entropy. The act of computation captures existing information, rearranges its symbols and produces an output or signal that we interpret as reality. Computation can describe all things, even the seemingly ephemeral. And, in turn, all things have the ability or potential to compute. A computer can be made of anything, even water (our brains) or sticks that fall from the trees on our front lawn!

Ed Fredkin believes in a "digital physics" -- that all the particles of the sub-atomic world are made of binary units of information and that the universe is governed by a single programming rule. The rule must be simple, and yet through its constant repetition it successively transforms simple bits of information and ultimately generates the incredible complexity of existence. All of the physical and biochemical processes of nature are controlled by the manipulation of information in its digital form. Fredkin's reasoning draws on the work of Hungarian mathematician and scientist, John von Neumann and his cellular automata.[441] Perfectly rational scientists believe that things just happen because they happen. But according to information theory, the structure of our world depends on patterns that determine what happens, not on substance.

---

[439] John Archibald Wheeler, "God is in the Machine," Wired, Kevin Kelly, Issue 10:12, 12/2002, from a 1989 Wheeler lecture.

[440] Ibid.

[441] As alluded to earlier, a cellular automaton is a system of cells in which each cell is in one of a finite number of states. Each cell transitions from one state to the next according to a rule where the outcome for a particular cell depends on the states of cells in the neighborhood of that cell. This can encompass every kind of discrete cellular process.

For example, we can think of an electron as a pattern of information and an orbiting electron as that pattern in motion. Imagine something like the home team's cheering section in the stands at a ball game -- each student holding a card that when held and displayed with those of his neighbors spells out or depicts a message. This is a kind of "von Neumann neighborhood." The bits of information never move, they just blink on and off, displaying different patterns. The configuration of the bits (those that are blinking at any given time) determines what the object or action is. Traditional physics assumes that reality is continuous, like a painting -- that one point in time flows into the next and that space is perfectly smooth. However, many things once thought to be continuous are now known to be discrete: the atom, the electron (a charge is discrete or quantized), photons (particles of light) and electromagnetic forces. Digital physics portrays created things as grainy in consistency, more like mosaics than paintings. This is an astounding conclusion. It is as if our lives play out like an old fashioned Nickelodeon, one frame at a time. And that is a conclusion that is very disorienting for most of us.

"As the consequences of the Quantum Theory became better understood, it became clear that the angular momentum of particles can only exist in multiples of +1/2 units of spin. This has the amazing consequence that a flywheel cannot have a continuous range of angular momenta, rather it must only have multiples of +-1/2h. Angular momentum is now known to be discrete. The story goes on with phonons and vibrons as quantized units of sound and other forms of energy."[442] Fredkin concludes that there is no convincing argument based on experimental evidence that shows any physical quanity to be continuous.

Of course, the bigger things get the more they appear to be continuous and not grainy. So, it is logical to conclude that the basic units of time and space must be imperceptibly small.

Time and space are discrete, and Fredkin concludes that the state of every point in space at any point in time is determined by a single mathematical relationship or algorithm. In this case, he is talking about a "recursive algorithm" -- an algorithm whose output is fed back into it

---

[442] "A New Cosmogony," Edward Fredkin, www.leptonica.com/cachedpages/fredkin-cosmogony.html

as input. This algorithm is instrumental in manufacturing the ultimate reality, what we see around us as our material reality.

The universe begins with simple bits of information and simple rules. The recursive algorithm (which keeps feeding itself and spitting out progressively more complicated outputs) results in the complexity that we are and that we experience.

But, if the universe is itself some kind of computer, what if it is a "reversible" computer?

What does that mean? We commonly think of computers as being irreversible. They compute and use the energy consumed in their computations in order to forget things, rather than to perform other functions – the energy used in computation disappears into the environment as waste. By contrast, a reversible computer would use no energy at all and have its history always available to be rediscovered. It would, in effect, be time independent. Its history would always be available to the operator. As a result, it would be something of a divine computer, independent of the limitations of time.[443]

And the idea of a reversible computer leads Fredkin to argue that analysis and reason lead one to the conclusion that our "something" has to have been initiated by a "something" beyond our "something." This is what he calls "other."

"If space and time and matter and energy are all a consequence of the informational process running on the Ultimate Computer, then everything in our universe is represented by that informational process. The place where the computer is, the engine that runs that process, we choose to call 'other . . .'" ". . . Computation does not require conservation laws or symmetries. . . does not have to have time as we know it, there is no need for beginnings and endings. . . (it) is compatible with worlds where something can come from nothing . . . the questions that

---

[443] Computation is not inherently irreversible – so you can theoretically build a computer that doesn't use up energy or give off heat, though none currently exists. Can one design a universal computer that doesn't dissipate heat? It would be logically reversible – a universal computer whose computational history can always be unearthed. If the movement of molecules and atoms is theoretically reversible, then it should be possible to reconstruct history by tracing the motion of microscopic particles backward. The theoretical feasibility of a reversible computer supports that claim. Fredkin proposes that a cellular automaton underlies reversible reality.

puzzle us about the origin of our universe are most unlikely to apply to Other... the problem of the origin of Other is tautologically null. Other is that place that has such structure and laws as to not raise the question of its origins, as origins are a concept peculiar to our world."[444] God, by any other name?

In Fredkin's mind, the rule is the prime mover. "What I'm saying is that at the most basic level of complexity an information process runs what we think of as physics. At the much higher level of complexity, life, DNA—you know, the biochemical functions—are controlled by a digital information process. Then at another level, our thought processes are basically information processing."[445] The ultimate computer, Fredkin says, is not in our universe. The hardware is beyond the grasp of the software. The conservation of energy says something cannot come from nothing. But Fredkin maintains that perhaps a different cellular automaton, with different rules created ours. This is, in effect, the Kabbalistic description of creation.

Fredkin sums up his speculation: "I find the supporting evidence for my beliefs in ten thousand different places. And to me it's just totally overwhelming. It's like there's an animal I want to find. I've found his footprints. I've found his droppings. I've found the half-chewed food. I find pieces of his fur, and so on. In every case it fits one kind of animal, and it's not like any animal anyone's ever seen. People say, where is this animal? I say, well, he was here, he's about this big, this that and the other. And I know a thousand things about him. I don't have him in hand, but I know he's there."[446]

GOD, INFORMATION, ENTROPY AND THE HOLOGRAPHIC PRINCIPLE

In the absence of a conclusive argument to the contrary, there must have been a starting point for everything that we recognize as reality.

---

[444] "A New Cosmogony," Edward Fredkin, www.leptonica.com/cachedpages/fredkin-cosmogony.html

[445] *The Atlantic*, "Did the Universe Just Happen," Robert Wright, April, 1998, p. 29-44

[446] Ibid.

The Elements of Jewish Exceptionalism

Science and "mysticism" seem to agree that information is is the stuff on which that foundation stands.

"'In the part lies the whole,' is a famous statement made by the great historian of Gothic architecture, Max Panofsky. In any given part of a Gothic structure, one can find the whole of Gothic design and art. The same is true for any great system of belief or thought. Its ideas and beliefs are coherent and consistent from setting to setting, from one part to another part."[447]

The Kabbalistic tradition asks mankind to utilize "spiritual" powers that are evidence of the information-based existence of which we are part. As in Gothic architecture, that information may well be expressed in a holographic model of our physical reality.

For thousands of years, the Chinese have practiced a form of healing that is a product of the information linking our organs and skeletal systems to specific points of contact on the surface of our bodies. Over 1000 such critical points of contact have been identified and organized by "meridians." Perhaps even more interesting is the belief by experts in this field that the image of a human appears in each ear, with points on a miniature fetus-like body that mirror those larger points along the body's meridians. That little man in the ear is used to diagnose problems elsewhere in the body. As summarized by Michael Talbot, "just as every portion of a hologram contains the image of the whole, every portion of the body may also contain the image of the whole."[448] Diagnosing or treating any particular part of the body through the corresponding holographic representation of that part of the body has been part of the science of acupuncture since its inception.

The manipulation of information innate to acupuncture is not limited to Chinese medicine. Part of the unique heritage of the Jews is the biblical command to wear phylacteries (*tefillin*) on one's arm and forehead. The oral tradition places those boxes and straps in a way that allows them to exert pressure on the very points on the head and arm that acupuncture associates with the treatment of mental clarity. "It appears that the *tefillin* and wraps form a potent acupuncture point formula

---

[447] "The Song of Chana: A review of the prayerful life of Chana in Sefer Shmuel," Rabbi Yechiel Poupko, September, 2014.

[448] *The Holographic Universe*, p. 115

focused on the Governing Vessel (Du Mai) and aimed at elevating the spirit and clearing the mind."[449] Steven Schram concludes: "If someone handed an acupuncturist the above point formula (showing the contact points of the *teffilin*) and asked what was being treated, there is little doubt that mental and shen issues would be a strong part of the pattern. What is surprising is that such a point formula would be found in a non-Chinese procedure that has been continuously practiced for many thousands of years. It may be that the originators of the *tefillin* ritual had some inkling of its special effects, even though they may have lacked the depth and specific knowledge we have today . . . Regardless of the belief system behind the procedure, it seems clear that putting on *tefillin* is a unique way of stimulating a very precise set of acupuncture points that appears designed to clear the mind and harmonize the spirit."[450]

---

[449] *Journal of Chinese Medicine,* "Tefillin: An Ancient Acupuncture Point Prescription for Mental Clarity," Steven Schram, #70, October, 2002, p. 5-8

[450] The command to wear what became tefillin is repeated four times in the Torah -- in Exodus 13:19 and 13:16 as well as in Deuteronomy 6:8 and 11:18. The Karaites, Philo and Rashbam on Exodus 13:9 thought it was a figurative command. Josephus in *Antiquities*, Book IV, Chapter VIII, 13 speaks about actual objects, "a remembrance on the arms." But in a note, William Whiston (translator) asks whether God meant for such Jewish "memorials" to be literally understood. The Letter of Aristeas (Hellenic, 2nd century BCE—translated by Avraham Kahane, Jerusalem, 1969, p. 49) talks about commanding the sign around the hand. The New Testament mentions the Pharisees and their tefillin, mentioning only the forehead (Matthew 23:5). Greek translations appear to refer to an amulet that the Pharisees tied to their foreheads. In fact the Greek word phylakterion means amulet.

TB Menachot 43B refers to tefillin on the head and forearm as an indicator of purity. There is a story of Miriai (as recorded in the Mandaean, pre-Islam gnostic, Book of John) about a Jewish woman who rejected Judaism and its customs, including totiftha, instead choosing a "headband." The same story speaks of the synagogue and of men and women engaged in study there. Does it refer to women wearing tefillin? Wearing actual tefillin was widespread in second temple times. Discoveries at Qumran and the Bar Kochba caves may mean that some, but not all wore tefillin. TB Shabbat 118b and the Talmud Yerushalmi Berakhot show that this may have been a custom of only the Hassidim. In his commentary on Parashat Bo, Saadiah Goan notes laxity with this commandment. The TB says that Sadducees wore it between the eyes, not on the forehead.

J.I. Packer & M.C. Tenney in their *Illustrated Manners and Customs of the Bible*, p. 482, and Saul Lieberman in his *Hellenism in Jewish Palestine*, p. 108, note 50, concur that this common practice arose to counter the plethora of idolatrous charms that were being worn. There is also some evidence that ancient Egyptians wore amulets on their foreheads and arms (Jamieson, Fausset and Brown, *Commentary on the Old*

Award winning physicist Jacob Beckenstein, developer of theories of black hole entropy, black hole thermodynamics and the Beckenstein Bound (similar to the holographic principle), summarized increasingly popular work done by John Archibald Wheeler.[451] It suggests that scientists may "regard the physical world as made of information, with energy and matter as incidentals." Quoting William Blake, he asks if the holographic principle implies that seeing the world in a grain of sand could be more than just poetic license.[452]

As a building block of creation, information is also understood to be indestructible. Once it is created, it is eternal. This is, in fact, an argument against Hawking's conclusion that black holes must destroy information.

The Holographic Principle represents that when an object is sucked into a black hole, its "substance" may be lost, but the information that defines it is retained, somehow imprinted on the surface of the black hole. The event horizon of that black hole (the border beyond which the object is irretrievable) retains the information of everything that crosses it. And what that tells us is that the information that comprises the three dimensional world in which we live is retained on two dimensional surfaces.[453]

Physicist Leonard Susskind says, "The information in a black hole is all on the surface of the black hole. So the more and more refined description you make of a system, you will wind up placing the information at a boundary. There are two descriptions of reality: either reality is

---

and New Testaments, p. 638 on Exodus 13:9). Rabbi Akiva commented that the word totafot means two and two in the languages of Katpi and Afriki (TB Sanhedrin 4B).

[451] Archibald Wheeler worked with Neils Bohr on nuclear fission, was a principal in the further development of Einstein's gravitational theory in the 1950s, coined the terms black hole and wormhole, and worked to develop a unified theory of quantum gravity. He was the mentor of Richard Feynman.

[452] *Scientific American*, "Information in the Holographic Universe," Jacob D. Beckenstein, 4/1/2007

[453] The holographic principle states that the physics of a 4 dimensional universe including gravity is mathematically equivalent to the physics on its 3 dimensional boundary without gravity. This implies that the reality that we see is just a holographic projection of information from the edge of our reality. This principle is present in String Theory and almost all approaches that attempt to unify relativity and quantum theory. (Hertog, Hawking, Hartle, arxiv.org/abs/1205.3807)

the bulk of spacetime surrounded by the boundary, or reality is the area of the boundary. So which description is real? There is no way to answer that. We can either think of an object as an object in the bulk space or think of it as a complicated, scrambled collection of information on the boundary that surrounds it. Not both. One or the other."[454]

Let's explain that idea a little more thoroughly. The event horizon of a black hole and the boundary of space-time are each analogous to holography's "light sheet." In explaining the holographic principle, one can compare the light sheet to a piece of film on which is imprinted all the information that comprises that object. But here, in Susskind's first conclusion, the light sheet precedes the image. It is not recording objects that already exist in the universe. Rather, it is projecting information from its surface that then becomes an object. So a two-dimensional information source projects our three dimensional reality. The light sheet may do more than simply generate all the forces and particles that we sense and experience. It may actually generate the substance of space-time itself. "'I believe that space-time is what we call emergent,' says Herman Verlinde, a physicist at Princeton University . . . 'It will come out of a bunch of 0's and 1's.'"[455] This is Susskind's "scrambled collection of information on the boundary." It is as if the universe is actually a two-dimensional canvas of information at the boundary of the cosmos.

Again, we encounter entropy, a product of computation that is best known as a characteristic of thermodynamics in the field of physics. Thermodynamic entropy describes changes that occur as a system becomes more disorderly while spontaneously evolving away from its initial (creative) conditions in accordance with the second law of thermodynamics.[456] Entropy is a measure of unpredictability in physics, but

---

[454] *Scientific American*, "Bad Boy of Physics," Peter Byrne, July 2011, p.83

[455] *Did the Universe Just Happen*, p. 35

[456] As was stated earlier, the second law of thermodynamics says that we live in a universe that becomes more disordered as time goes on and there is no way to change that. The world is degenerating. Even the most advanced machinery will waste energy and run down. It suggests that the universe will eventually exhaust its available energy and reach a point of stasis called "heat death." There are four laws of thermodynamics. The zeroth law of thermodynamics says that if two systems are in thermal equilibrium, each with a third system, then they are in thermal equilibrium with each other. The first law says that you cannot get something for nothing (energy

as was referenced earlier, it also describes information content (Shannon Entropy). The information content of an event depends on how probable the event is. The more surprising an event, the greater its information content. So information content is inversely proportional to probability.[457] By reasoning that both thermodynamic and Shannon entropy are equivalent expressions of a single creative principle, an arrow is pointed at the role played by information in the "creation" of matter – the very information referred to in the holographic principle.

As early as 1957, E.T. Jaynes[458] said that thermodynamic entropy in statistical mechanics is itself an application of Shannon's information theory. The thermodynamic entropy is understood to be proportional to the additional amount of Shannon information needed to define the detailed microscopic state of a system. That information is needed because it is not communicated by its macroscopic, classically thermodynamic description. In the case of a physical system, adding heat increases its thermodynamic entropy because it increases the number of possible microscopic states for the system, adding information and making a complete description of that state longer than it might otherwise be.

In the context of the holographic principle, keeping all information in the two dimensions of the "light sheet," tells us that the entropy of ordinary matter is proportional to surface area, not to volume. Three dimensional volume does not really exist because the universe is really a hologram originating with the information inscribed on the two

---

cannot be either created or destroyed so the total energy of the universe remains the same). The second says you can't even get something for something (the entropy of an isolated system that is not in equilibrium will increase over time and approach its maximum at equilibrium – changes are irreversible; a system cannot return to its original state without exacting a price from its surroundings). And the third says that as the temperature approaches absolute zero, the entropy of a system approaches a constant minimum. Classical thermodynamics is limited to equilibrium situations. We imagine that the system is always in equilibrium, even if the equilibrium shifts from moment to moment. Quantitatively, entropy is the amount of heat exchanged in a process divided by the temperature. In an isolated system, entropy always stays the same or increases.

[457] In information theory when there is a binary choice and the probability of either outcome is equal, the entropy is maximized. It is simply not possible to accurately predict the outcome – which bit of information will be chosen -- beforehand.

[458] E.T. Jaynes wrote extensively on probability theory and was a student of the physicist Eugene Wigner. His work on Bayesian thinking took him into information theory.

dimensional surface of its boundary.[459] A hologram is a "virtual" image. It appears to have a location, but has none. Instead the actual location of the hologram is, if you will, in the surface of the "film" that records it or contains its information. In religious terms, the home of divine information has no location but projects itself into what becomes our reality – the reality of our bodies, the objects around us, and even the immateriality of our minds.

CIRCUMCISION, ENTROPY AND JEWISH EXCEPTIONALISM

Let's continue with this analysis of the relationship between information and the divine, and while doing so, tie it to the chosen concept that initiated this discussion.

The thermodynamic concept of entropy to which we continually return, can help crystallize just how one's mindset must be altered to understand the relationship between faith and science. Can we identify a scriptural commandment that physically depicts Jewish exceptional status while leading directly and very logically to a theory that makes the physical and spiritual indistinguishable from one another?

Perhaps more than any other purely scientific notion, the concept of entropy provides substantiation for law, action and language that we routinely associate with only the spiritual and the divine. The earlier reference to the Hebrew word KR and its association with both chance and cold demonstrate just how essential scientific fundamentals are to understanding Hebrew scriptural narrative and its depth. Entropy explains the relationship between seemingly disparate Torah thoughts, their physical manifestations, and the heretofore incomprehensibility of their actual meaning. Working with this "entropy analog" we can get a strong sense of both why and how, as in the creation story, God chose to present his works and dictates in "the language of man" or in orally transmitted imagery rather than in the science and mathematics on which they are built. To use the Kabbalistic euphemism, the essence of our existence is "clothed" in our religious terminology and our macroscopic actions.

---

[459] Efforts to prove that the universe is holographic include the applications of a gravitational wave detector of GEO600 by Craig Hogan, as well as holometers (laser interferometers -- the most precise clocks in the world).

The commandments concerning ritual impurity are specific to the chosen people and are an example of the parallelism of scripture and science. These laws are routinely dismissed as archaic, irrelevant to modern existence. In even the most strictly religious circles, the associated laws are practiced and studied as a spiritual tradition, homage to an irretrievable past, completely disassociated from the physical. These rituals are associated with death in some cases and with disease or gender in others.

Particularly vexing are the laws pertaining to the bodily emissions of both men and women, especially in association with reproduction. In the twelfth chapter of Leviticus instructions are given defining a period of time after the birth of newborn during which the mother is considered contaminated or impure. When the offspring is a boy, the mother remains contaminated for seven days and on the eighth day the baby undergoes circumcision. A second stage of maternal contamination lasting 33 days follows, for a total of 40 days of impurity. In the case of a baby girl, the mother is considered contaminated for 14 days, followed by a second stage of 66 days for a total of 80 days. Interestingly, as relates to the earlier discussed the Hebrew word KR, which in one sense means cold, the thirteenth century commentary on the biblical commandments, Sefer HaChinuch says, "However, the conception of a female indicates coldness in the woman's physical nature, and in coldness the excess substances increase."[460]

Can the difference in the periods of contamination based on the gender of the offspring really be explained scientifically? And why is the flow that accompanies birth considered contaminated or impure in the first place?

According to the scriptural text, at the conclusion of the period of impurity, the mother brings two animal offerings to the Tabernacle or Temple, one to praise God and the other to "atone" for the "error" associated with the birth. The Hebrew word *chatat* is generally defined as "sin," but more accurately refers to an unintentional "missing of the mark," so to speak. So the biblical requirement to bring this particular offering is an indication that the birthing process is not quite perfect. Is that imperfection the releasing of fluids that in a perfect world would not

---

[460] Sefer HaChinuch, Mitzvah 166 on Leviticus 12-13

exist at all? We refer to these "waste" fluids as impurities. The mother brings an offering to acknowledge them and, along with the passage of time, compensate for their effects.

The impurity associated with birth has many exegetical explanations, but most equate it with the uncleanliness attached to the *niddah*, the menstruating woman. Others refer to the birthing process as the equivalent of an illness. But the idea of entropy provides some insight into just what the process represents in the scheme of the divine construction of our reality.

Consider creative processes in general: all conclude with not just the output sought, but also with a form of waste arising from the process itself. Entropy measures that waste. Mechanical, chemical and biological processes are all descriptions of the performance of work, and each type of work creates waste in the form of heat. When the universe came into being from a single point of immeasurable density, it burst forth into a fireball generating such extreme heat that elements could not form for another 300,000 years, after the period called inflation. Only then had the universe sufficiently cooled to allow the formation of the elements. This is perhaps the most fundamental example of the waste or disorder that grows out of the creative process. Of course, our lives are replete with examples of waste heat, including everything from exercise to simple friction to the heat generated by the internal calculating activities of our computers (the creation of new forms of information).

And in generating their own waste products biological processes mimic the production of thermodynamic entropy.

Let's look again at Leviticus, chapter 12 and the case of the new mother. The biological process of creating a new life involves innumerable, discrete, chemical combinations – each a creative activity in its own right. While every one of these results in a new compound or substance, each also produces an effluent or waste that is superfluous to the post-creative process. The birth of the child releases the sum of these waste "flows" into our reality. But just as the waste heat from a creative process must be dispersed before that creation can be actualized, neither the child nor the mother can move beyond the creative act in the presence of these wastes. Thus, the Torah prescribes a period of time during which the body expels all of the waste, leaving itself in its ideal post-creative state.

This passing of time lays the groundwork for a "rectified" post-creative state for the mother. But what of the baby? If it is a female, then time addresses this matter. The waste is dissipated, just as is heat in the case of thermodynamic entropy, while the new creation begins to thrive. And the period specified in Leviticus provides the necessary time for the waste to be nullified. The end of that time period is marked by an act that is at once both physical and symbolic: the sacrificial service. The mother and baby are both purified by time, 80 days in this case – but the purification of the baby also includes the symbolic act of the sacrifice. And in the case of the male new-born? Circumcision is also at once both physical and symbolic. The excision of the foreskin represents both the actual removal of the remaining creative waste that clings to the infant (just as it does to every male mammal) and is symbolic of purification, while the passage of time performs this deed for the mother.

Leviticus 22:37 gives us more food for thought. "When an ox or sheep or goat is born it will stay with its mother for seven days, and from the eighth day, it will be acceptable as an offering by fire to God." What happens in those first seven days of life? Does the *tumah* or impurity associated with the creative act fall away after seven days? How is this different from the experience of the new-born child? Are we to believe that the circumcision of a baby boy is analogous to the sacrifice of an eight day old ox, sheep or goat? This apparent parallel has led many commentators to conclude that the excision of the foreskin is tantamount to a sacrificial offering. In fact, in the 16th and 17th centuries, prayers comparing the circumcision service to the sacrificial service became part of the circumcision ritual.[461]

But let's look again at the connection made above between the creative act and its waste. The waste must be "nullified" or rendered impotent by being physically removed from the Jewish life that its presence will taint. An adult female stands by while the offending effluent is naturally diminished and ultimately eliminated – exactly analogous to the dissipation of heat thermodynamically. The female human that is created sees the same process unfold over time for her and

---

[461] *Why Aren't Jewish Women Circumcised?* See p. 32 for the prayers said in this vein

the completion of that process is marked with a sacrificial recognition of divine participation. As with the animal, the impurity attached to the baby boy has dissipated after seven days. Each of these newborns is now free of impurity – with one exception. Both the new born male and his animal counterpart are born with a foreskin. This piece of extraneous skin is the additional waste from the creative process that must be eliminated by human effort. The Midrash refers to the foreskin as a "sore" or a "boil" that must be removed from the body.[462] The animal, by contrast, carries that "sign" with him for the rest of his life, and, for him it serves a constructive function, harboring certain beneficial bacteria.

The Jewish male excises that waste as a sign of his exceptional status.[463] He is distinguished from the animals as well as from other people. Both this action and that of the mother in making a sacrificial offering mark the special status of the Jewish offspring, both male and female – a status that is enabled by the removal of "death" or waste at the beginning of life. Both male and female become equally "cleansed" by this process, for the fact that the uncircumcised may not eat of the Passover sacrifice applies only to the males. The females are prepared by the birthing process alone to do so.

The circumcision as a stage of the birthing process is associated with purification in other ancient societies as well. Herodotus noted that this was the case for Egyptians who preferred to be pure, even at the expense of reducing their physical beauty.[464]

While the eight day old animal is now a candidate for sacrifice, the eight day old Jewish male baby is a candidate for life among a "kingdom of priests." Whether he achieves that status or not is a product of his own effort as well as that of his family and community.

This unusual action or sign, acknowledging the chosen status, did not go unnoticed by supercessionists. In disseminating what was to become Christian theology, Paul insisted that circumcision should be anathema to God's faithful followers. He pointed to Abraham as being

---

[462] Genesis Rabbah, 46:10 467

[463] Genesis 17:11. "You shall circumcise the flesh of your foreskin, and that shall be the sign of the covenant between me and you." This is understood to be a sign by which God remembers the covenant that He made with Abraham and his offspring.

[464] Herodotus 2.37.2

worthy of special note even before he was circumcised.[465] "Nor is true circumcision something external and physical. He is a Jew who is one inwardly, and real circumcision is a matter of the heart, spiritual and not literal."[466] Christianity began to view the circumcision as a right of entry into not just the covenant between the Jews and God, but into Judaism itself.[467] They drew a direct parallel between it and baptism. To them it was a sacrament. The entropy analog makes the fallacy of this approach obvious. The Christian assertion that the non-circumcision of women is an argument against the physical procedure of circumcision itself is similarly contradicted by this "thermodynamic" explanation.

WHERE DO CONSCIOUSNESS, MIND & SOUL FIT IN?

If the text of the Torah is the word of God, and the text has designated one particular people as chosen, then shouldn't there be a detectible physical, mental or metaphysical difference between the Jew and his fellow man?

Both the mystical and rabbinic traditions allow that the Jewish soul may be at the heart of that difference. However, the great medieval exegete Maimonides, disagreed.[468] He believed that while the Jew was superior, that status could be attributed to a unique set of divine values and experiences. The *Zohar*, on the other hand offered that the Jewish soul emanated from a holy source while all others arose from "impure sources and render impure anything that approaches them."[469] This holier, higher level of Jewish existence was also integral to the understanding of many legendary rabbis over the last millennium including Yehudah haLevi in his *Kuzari*, the Maharal of Prague (Yehudah Lowe)

---

[465] Romans, 4:3

[466] Romans 2: 28-29

[467] *Why Aren't Jewish Women Circumcised?* P.83, "With the death and resurrection of Christ, however, the old Israel was replaced by the new Israel, the old covenant by the new covenant. Jewish particularity no longer had any reason to continue, since all of God's children were now to be brought into the universal church. Ethnic Israel, Jewish particularity, and Jewish circumcision were all supposed to come to an end. Baptism replaced circumcision and Christianity replaced Judaism."

[468] *Maimonides Confrontation with Mysticism*, Chapter 7

[469] Zohar, Genesis, #170, Sulam Edition

in his *Tiferet Yisrael*, Chaim Vital (who explained that every Jew possessed two souls, one attached to God), the Ramchal (Moshe Chaim Luzzatto who believed that Jews and non-Jews were effectively of entirely different species[470]), and the Ba'al Tanya, Rabbi Shneur Zalman, who wrote that the soul of the Jew is a part of God.[471]

In Kabbalistic thought, the soul is not distinguishable from either human consciousness or the mind. The traditional mystical belief, based on the *Zohar* itself, is that the human soul has five distinct levels, ranging from the mundane to the exceptional. And the exceptional is attainable by the Jew alone. The Nefesh is considered to be the lowest level or part of the soul and is associated with physical or animal-like instincts and desires. The Ruach (or spirit) provides its host with the ability to distinguish between good and evil. The Neshama allows man to realize his intellect. It is believed that this part of the soul is eternal. The exceptional parts of the soul, the Chaya and the Yechida are discussed in the section of the *Zohar* called *Raaya Mehemna*. The Yechida is available to only a few. These aspects of the soul are not associated with the body at all. The Chaya allows man to become conscious of the divine forces in life, while the Yechida actually represents a union of sorts with God. As understood by the Kabbalah scholar Gershom Scholem, "The Nefesh remains for a while in the grave, brooding over the body; the ruach ascends to the terrestrial paradise in accordance with its merits; and the Neshama goes directly back to its native home (the Divine world or Divine Mind, the *sefirah Binah*)."[472]

The soul/mind/consciousness is decidedly immaterial and exists in the absence of the body. And while all living creatures have a soul, the Jewish soul, according to the *Zohar*, has some higher level components.

What then is the relationship between information, the building block of all substance, and this aspect of humanity that seems equally immaterial?

Is consciousness a function of or a product of the brain? Is that where it resides? Mathematician John von Neumann calculated that

---

[470] Dereck Hashem, 4:1

[471] These remarks are taken from "The Soul of a Jew and the Soul of a Non-Jew," Hanan Balk, p. 47, Hakirah.org

[472] *Kabbalah*, p. 161

during the average lifetime of a human being, the brain will store 2.8 times 10 to the 20 bits of information. Brain research has failed to account for just how this can be physically possible. So information must be pervasive and its "location" unidentifiable – existing everywhere and nowhere at once.

The ancient Egyptians believed that the soul was found in the heart. Leonardo da Vinci attempted to find it by dissecting the brain. An American physician, Duncan MacDougall concluded in 1907 that the soul weighed 21 grams. The Jewish oral tradition that is associated with sacrificial rites locates the soul in the blood. Since the search for a physical soul seems to lead to a dead end, its association with consciousness and the mind has become common. "The 17th century French thinker Rene Descartes proposed an influential theory that leaned on neuroanatomy as well as philosophical inference. He declared the pineal gland, a pea-size glob just behind the thalamus, the seat of consciousness, 'the place in which all our thoughts are formed.' But Descartes was a dualist: he believed that body and mind are separate and distinct. Within the physical matter of the pineal gland, he reasoned, something inexplicable must lie, something intangible – something that he identified as the soul."[473]

Bertrand Russell once said that consciousness provides a window into the brain. While many philosophers and scientists believe the brain to be the home of both mind and consciousness, what they actually are, how they arise and how they interact with nature is increasingly being examined in quantum, scientific terms. Consciousness is often linked to quantum physics and simultaneously to information theory. If bits of information are actually the building blocks of creation, then consciousness and the mind, and the brain for that matter, are products of those same building blocks. Descartes' dualism is necessarily reduced to monism – not the material, biological and chemical monism of the traditional reductionist, but the monism of information.

The conversion of information is increasingly understood to be the basis of both the physical and spiritual. Similarly the principles of two information-based features of mathematics and science, holography and entropy, are windows into the immateriality of the mind. Neither

---

[473] *The Atlantic*, "Awakening" by Joshua Lang, Jan/Feb 2013, p. 54

Dualist nor reductionist theories are applicable to an information-based existence.

Consciousness is classically understood to be the awareness of oneself and one's surroundings. Research in the field rests on the chemical and biological activities of the brain. In 1994 David Chalmers addressed the problem with this approach: He explained that problems with the examination of consciousness could be categorized as either "easy" problems or "hard" problems. Easy problems are solved by determining the mechanisms that explain behaviors. They utilize psychological, cognitive and neuroscientific analysis. The hard problem is what he calls "the explanatory gap" -- determining how simple physical and biological processes "translate into the singular mystery of subjective experience. If this gap cannot be bridged, then consciousness must be informed by some sort of inexplicable, intangible element. And all of a sudden we are back to Descartes."[474] "The hard problem is explaining how subjective experience arises from neural computation. The problem is hard because no one knows what a solution might look like or even whether it is a genuine scientific problem in the first place. And not surprisingly, everyone agrees that the hard problem (if it is a problem) remains a mystery."[475]

Some want to believe that the hard problem is really not that hard at all: ". . . the feature (of the easy and hard problems) that (scientists) find least controversial is the one that many people outside the field find the most shocking . . . the idea that our thoughts, sensations, joys and aches consist entirely of physiological activity in the tissues of the brain. Consciousness does not reside in an ethereal soul that uses the brain like a PDA; consciousness is the activity of the brain."[476] This approach follows classical physics -- the science of matter -- and consciousness arises from matter. In classical physics, systems are made of independent parts that interact with their neighbors and generate deterministic behaviors. So, one must conclude that consciousness is a product of a physical system -- the brain.

---

[474] *The Atlantic*, "Awakening" by Joshua Lang, Jan/Feb 2013, p. 54

[475] *Time Magazine,* "The Mystery of Consciousness," Steven Pinker, 1/29/2007, p. 61

[476] Ibid, p. 62

Neither reductionists nor dualists are about to give up without a fight. David Chalmers has introduced the concept of "fundamental property dualism." He explained that the universe has conscious properties just as it has physical properties. The conscious and physical properties interact causing transitions in physical states that may affect human consciousness. Some have gone a step further, proposing another form of dualism, Panpsychism, which assumes that all physical elements of reality have some consciousness associated with them – saying, in effect, that the universe is based on this consciousness.

Quantum physicists have entered the debate suggesting that our knowledge of the quantum world can change our understanding and analysis of consciousness. Until recently, neurological studies of the brain and consciousness have taken classical physical relationships as a given. Quantum effects, if known, were ignored. We have now learned, however, that electrochemical analyses of brain function are not able to explain how distant entities become "entangled" (in the language of quantum mechanics). But information can. We spoke about entanglement earlier. This is a feature of something called Bell's theorem, which showed that everything is continually and permanently interacting.[477]

Physicist David Bohm believed that the entanglement of sub-atomic particles, even at great distances, was the product of the illusion of "separateness." At the deepest level of creation, entities "are actually extensions of the same fundamental something."[478] We are calling that "something" information. In 1927, Werner Heisenberg (originator of his eponymous uncertainty principle) showed that there is an underlying unpredictability in nature, even though Newton had earlier ruled that out.[479] And, many now see that mind/body dualism strongly correlates

---

[477] In 1964, John Bell developed a mathematical proof that showed the possibility of quantum correlations arising from other than just local causes. There can be correlations between experiments in two widely distant laboratories, for example. This quantum "entanglement" seems to indicate that reality is "non-local," meaning in Bell's words, "the setting of one measuring device can influence the reading of another instrument, however remote."

[478] *The Holographic Universe*, p. xv

[479] Heisenberg's Uncertainty Principle states that the more precisely the position (of a particle) is determined, the less precisely the momentum is known in that instant, and vice versa.

with the wave/particle dualism of the quantum world – the very definition of unpredictability. Quantum theory leads us to the understanding that matter is not, in fact, solid, nor can it be differentiated from energy, and ultimately, information. And this seems to be the same "substance" from which our souls and consciousness are formed.

QUANTUM CONSCIOUSNESS

If consciousness is not physical in the classical sense, then isn't it all the more likely that the mind and soul are based on information alone? Is there any other solution that does not involve a biological, chemical or physical source? These once outlandish ideas are making inroads among researchers and academics.

One of the earlier theorists was Alfred Lotka, a Ukrainian chemist who posited in 1924 that the mind controls the brain by controlling the quantum jumps that would otherwise lead to complete randomness.

Nick Herbert[480] not only declared that consciousness is everywhere and in all things (panpsychism) but is determined by quantum effects. He theorized in Copenhagen style that objects acquire attributes only after they are observed. This is tied to the randomness and interconnectedness that John Bell discovered: once two sub-atomic particles have interacted they remain connected (entangled). Herbert assigned randomness to free will and Bell's interconnectedness to deep psychic connections.

Evan Walker's synaptic tunneling model[481] declared that electrons tunnel between adjacent neurons creating a virtual neural network overlapping the physical one. Thus, a virtual nervous system directs the behavior of the physical nervous system. These electrons pass through energy barriers that they should not be able to climb according to classical theory. Walker concluded that the brain's behavior can be described by classical laws while consciousness can only be described by quantum laws.

---

[480] American physicist and author of *Quantum Reality: Beyond the New Physics*, NY, Doubleday, 1985

[481] *The Physical Nature of Consciousness*

Eugene Wigner (Hungarian physicist) proposed adding a term to Schroedinger's equation[482] (the one that describes an observer's presence collapsing the wave and crystalizing an event) that makes it non-linear and explains the collapse of the wave. This term is a measure of information which disappears once a measurement is performed (indicating the presence of an observer).

John C. Eccles[483] offered that the mind exists outside the material realm and interacts with the brain at the quantum level. These immaterial "particles" that he calls psychons, cause voluntary excitations of neurons.

Michael Lockwood posited that an immaterial consciousness scans the brain, finding existing sensations which are physical attributes of the brain. It then takes on the role of the quantum observer, and by doing so (collapses the wave, in Schroedinger terms) creates a physical reality.

These theories increasingly state or imply that there are material mechanisms that organize and manipulate information, connecting an immaterial consciousness to the physical brain.

It's not necessary to understand what a Bose-Einstein Condensate (achieved experimentally in a gas in 1995 after being theorized in 1925) is to get the point of the next few paragraphs. But pay attention to the conclusion:

Bose-Einstein Condensates are the most highly ordered structures in nature (surpassing crystals). Each of their constituents appears to occupy all of their space and all of their time. All the constituents share the same identity. Together they behave like a single constituent or like a single particle. This means that all the atoms in the condensate behave identically. As temperature drops the wave of each atom grows until the waves of all the atoms begin to overlap and eventually merge. They

---

[482] Schroedinger's equation (1926) describes fundamental quantum mechanical behavior. It is a partial differential equation that describes how the wave function of a physical system evolves over time. This differs from classical physics in which Newton's second law describes motion (F=ma) and is used to predict what a system will do after it moves away from its initial conditions.

[483] "Do Mental Events Cause Neural Events Analogously to the Probability Fields of Quantum Mechanics?" *Proceedings of the Royal Society of London, Series B. Biological Sciences*, Vol. 227, No. 1249, May 22, 1986, pp. 411-428.

then become indistinguishable from one another, completely losing their individuality.

What if living things could achieve this result? Heisenberg's uncertainty principle makes it impossible for atoms that have reached the lowest possible energy to reach zero energy. This is instead called "zero-point energy"-- the minimum amount of energy that an atom can have. Herbert Froehlich proved the feasibility of Bose Einstein condensation at body temperatures (rather than in extreme cold) in living matter (in cell membranes). If this is so, then all living systems could contain B-E condensates.

And if living things contain B-E condensates, then living things can act as quantum computers,[484] efficiently manipulating and storing information.

Electrically charged molecules of living tissue behave like electric dipoles (a pair of separate electric charges of equal magnitude but opposite charge). Digestion of food creates energy and all molecular dipoles line up and oscillate in a coordinated manner which may result in a B-E condensate. These biological oscillators, like laser light, can amplify signals and encode information (store external stimuli). This is crucial to integrating information in the brain. The molecules in each cell may be doing a kind of quantum computing using associative memory.

Are you lost? Try to refocus now, because here is the salient question: Is there a predisposition for this biological/electrical manipulation of information to work for the benefit of mankind? Is the process biased toward human survival?

Ian Marshall[485] speculated that when the condensate in the brain is excited by an electrical field, conscious experience occurs. The brain

---

[484] Contrary to our understanding of computing, in quantum computing the bit does not take a definite value of 1 or 0. Instead, a quantum qubit can exist simultaneously as both 0 and 1or in any combination of the two binary states. And that is because quantum particles can exist in two locations or states simultaneously. That is called superposition. The qubits can be connected to each other through entanglement even though they are spatially separated. Because an action performed on one effects the other, quantum computers have incredible parallel processing ability. A classical computer can handle only one possibility at a time. A quantum computer can test every possible solution at once.

[485] *New Ideas in Psychology*, "Consciousness and Bose Einstein Condensates," I.N. Marshall, 1989, Vol. 7(1), p. 73-83

The Elements of Jewish Exceptionalism

has the ability to organize millions of neural processes into a coherent whole of thought thanks to an underlying quantum "coherent state," which is the B-E condensate.

Marshall's work has made him a strong advocate of the Strong Anthropic Principle (which states that all of what is happening in nature is aimed at mankind). He concludes that the wave function tends to collapse toward B-E condensates – meaning that there is a universal and completely natural tendency toward creating the living and thinking structures that we see around us. The universe has an innate tendency toward life and consciousness.

Similarly, Marshall thinks that natural selection can't operate quickly enough to get rid of all the mutations that arise in nature,[486] so another force must select the mutations that are most useful for survival. That would be the wave function's tendency to choose states of life and consciousness – a predisposition for the world we have come to know.

Now, let's return to consciousness, what it is and where it can it be found. Reductionists have suggested everything from "the anterior cingulate cortex, a region also associated with motivation, to some parts of the visual cortex, to the cytoskeleton structure of neurons. . . Some theories peg consciousness not to a particular part of the brain but to a particular process, such as the rhythmic activation of neurons between the thalamus and the cortex."[487] The resistance of the academic community to "giving up the ghost" is startling.

". . . there is growing interest among the scientific community in String Theory, the only branch of mathematical geometry that has successfully explained recent discoveries in high energy particle physics. The mathematical proof of string theory requires the existence of a ten dimensional universe. But in trying to map consciousness, science continues to limit its data search within time and the three spatial dimensions, ignoring six of string theory's ten dimensions. Perhaps because mainstream science tacitly limits itself to four of the ten dimensions, progress has come to a dead end. Accordingly, the most widely accepted theory among neurophysiologists is that consciousness is somehow a

---

[486] See the work of John Cairns in the 1980s concerning adaptive mutations

[487] *Scientific American*, 11/2009, p. 25

'byproduct' of the electrical sparking of neurons, an accidental epiphenomenon of the activities of nerve-filled wet meat."[488]

David Chalmers sums up the frustration that accompanies contemporary scientific research about consciousness: "The hard problem is that of experience: why does all this processing give rise to an experienced inner life at all? Even Crick and Penrose concede that so far they have little idea how the problem might be solved. They simply hope that if we do enough investigation in neuroscience (for Crick) or in physics (for Penrose), the faint outlines of a solution might be revealed. At this point, such remarks are not much more than an expression of faith."[489] The irony of his use of the word "faith" should not be lost.

THE MUSIC OF THE UNIVERSAL SOUL

The mathematician and computer scientist, James Culbertson believes that every object has a degree of consciousness. He reasons that all space-time events are conscious of other space-time events. According to Culbertson, the experience of an event is static. Special circuits in our brains create the impression of time flow and time travel through the region of space-time events connected to the brain. Memory of an event is nothing more than re-experiencing its link to space-time. We don't store that event, we just link to it. It's something akin to an information link in cyberspace except that here, the Wi-Fi network or cloud is all-encompassing (the quantum field of physics, the collective unconscious, the morphogenetic field or Bohm's implicate order). Rather than randomness determining quantum jumps, it is space-time history that does. As physicist David Bohm viewed it, both relativity and quantum physics recognize that the universe has a flow that integrates everything.

---

[488] By S.R. Joyce, B.S. Electrical Engineering, M.A. Asian Philosophy, CIIS Doctoral Candidate, Department of Philosophy & Religion, Concentration in Philosophy, Cosmology, and Consciousness, PARP 9600 – Doctoral Comprehensive Exam Paper, Professor Allan Combs, School of Consciousness and Transformation, 2013

[489] *"Review of Journal of Consciousness Studies"*, David J. Chalmers, *The Times Literary Supplement*, October 1994, p. 48.

Bohm, a student of Albert Einstein, introduced the idea of the implicate order to the conscious world in the 1950s. He believed that we perceive the flow of reality through momentary, still images which are a kind of simplification of motion. Similarly, the flow or stream of consciousness in our mind allows us to extract momentary images, concepts, ideas and other forms of thought. In this framework, thought is a kind of movement and concepts are kinds of objects. The implicate order is a kind of higher dimension where there is no difference between mind and matter. The difference is only in the explicate order (the conventional space-time of physics).

Interestingly, Bohm believed that when we listen to music (pure information) we directly perceive the implicate order, not just the explicate order of the sounds.[490] Music is our most understandable bridge between information and materiality. It is mathematical information that manifests itself in physical vibrations that our senses feel, absorb and manipulate. Scholars and sages of antiquity were on the same page. In both the books of Samuel and Kings the prophetic state was brought on with music.[491] The information retained in the implicate order can be captured and assimilated by one who attains the prophetic state.

"How does a prophet go about getting prophecy? He must clear his mind by meditating. According to the Rambam, one of the ways to do this was to play music. Listening to music could be a type of meditation, especially if it is played a certain way . . . Rabbi Chaim Vital writes in the fourth part of *Sha'arey Kedusha* that those seeking prophecy would play music until they could get into a trance, at which point the music would stop . . . Music shares the same spiritual source as prophecy. It

---

[490] The "other worldliness" of music has been recognized since man's earliest days. In addition to music being the background for religious prophecy, studies show that infants possess sophisticated cognitive musical abilities and reasoning (DeCasper AJ, Carstens AA, "Contingencies of stimulation: effects on learning and emotion in neonates, Infant Behavior and Development," 1981, 4:19. Also, Krumhansl CL, Jusczyk PW. "Infants' Perception of Phrase Structure in Music." *Psychological Science*, 1990, 1:70. Also, Wynn K., "Addition and Subtraction by Human Infants," *Nature* 1992; 358:749) Leng and Shaw proposed that music is a "pre-language" available at an early age to access the brain's inherent firing patterns and enhance the ability to perform special-temporal reasoning (Leng, Shaw, "Toward a Neural Theory of Higher Brain Function Using Music as a Window," *Concepts Neuroscience* 1991; 3:239)

[491] I Samuel 10:15 and II Kings 3:15

therefore has the power to cut away and penetrate all barriers that prevent prophecy."[492]

In his seminal work, *The Book of Beliefs and Opinions*, tenth century sage Saadia Gaon remarks eight types of musical rhythm that affect human temperament and mood. He discusses how they should be mixed in order to lead men to the ideal balance in behavior.[493]

The Gemara refers to the heavenly music of both the angels and the righteous in several places,[494] while its principals emphasized that scripture and law were to be both read and studied with a melody: "Sing it every day, sing it every day,"[495] said Rabbi Akiva. Whoever reads (the Torah) or studies the Mishna without song may have the verse applied to him (from Ezekiel 20:25), "I also gave them statutes that were not good and laws whereby they should not live."[496]

King David addressed God as "the Conductor" in many of his psalms. He saw God as orchestrating the symphony of life. And in the times of the First and Second Temples which were the centers of spirituality in the first and second Jewish Commonwealths, music was understood to be an essential wisdom.

PANPSYCHISM AND ITS BELIEVERS

Consciousness and knowledge are fundamentally linked. Consciousness is the sum of knowledge and information enabled by awareness, while knowledge is acquired through consciousness.

Quantum theories and reflections on consciousness rest on the assumption that the quantum field contains active information that determines what kind of a particle emerges and what happens to it. That active information is the "proto-consciousness" of the particle. Every particle has this elementary consciousness -- so all matter has mental

---

[492] *Innerspace,* p. 149 and 225. See also his notes for various sources in Tanach, Zohar and the commentaries.

[493] *Emunot V'Deot*, Saadia Gaon, Section 10, Chapter 18

[494] See TB Chagiga 12b and 14a, Avoda Zara 3b, Eruvin 21a, Sanhedrin 91b, Megilah 10b

[495] TB Sanhedrin 99a

[496] TB Megilah 32a

properties. Information is the bridge. Particles are made of information, one arises from the field from which the other is the substance. Bohm theorized that at the level of the quantum field or implicate order, consciousness and the physical are the same.

And Bohm has lots of company. Panpsychism (mentioned earlier) is the notion that everything is conscious to some extent. It is a feature of the philosophies of Plato, Spinoza, Leibniz and William James. Said Plato, "This world is indeed a living being endowed with a soul and intelligence . . . a single visible living entity, containing all other living entities, which by their nature are all related."[497] Even neuroscientist, Christof Koch, who has dedicated his research to the physical examination of consciousness, has offered a scientifically refined and immaterial version of this philosophy. Koch believes that consciousness arises within a sufficiently complex information processing system. He believes that even the internet could be conscious because that is the way the universe works. "Likewise, I argue that we live in a universe of space, time, mass, energy and consciousness arising out of complex systems."[498]

Philosopher Thomas Nagel is a vocal opponent of reductionism and offers a variation on this theme, concluding that protomental properties must be present in all matter, and when organized become consciousness. He believes in one, common source for the material and mental aspects of the world.[499] Fellow philosopher Robert Adams proposes ". . . that everything that is real in the last analysis is sufficiently spiritual in character to be aptly conceived on the model of our own minds, as experienced from the inside. . . Now if it is indeed right that things in themselves must have intrinsic non-formal qualities, and that such qualities must be conceived as qualities of consciousness or analogous to qualities of consciousness, it follows that things in themselves must be conceived as all having qualities of consciousness or qualities analogous to qualities of consciousness. And that is at least very close to

---

[497] Plato, Timaeus, 29/30

[498] *Wired.com*, Brandon Keim, 11/14/13, "A Neuroscientist's Radical Theory of How Networks Become Conscious."

[499] *Mind and Cosmos*

the conclusion that things in themselves must be conceived as having a spiritual or mental or at least a quasi-mental character." [500]

Panpsychism has a home in ancient and medieval religious thought as well. And in that regard, both science and philosophy find common ground with Kabbalistic thought. In Kabbalistic terms, this information based idea of consciousness is related to the *sephira* called *Daat*. *Daat* is not technically one of the *sephirot*, but is the location in the "Tree of Life" of Kabbalistic imagery where all the others are found. In *Daat* (which means knowledge) all of the *sephirot* share equally. The soul, or this "proto-consciousness," can be physically activated by information (*Daat*) which gives it form.

Similarly, but reflecting decidedly non-Kabbalistic thought, Maimonides concluded that all objects have a "soul" (or for our purposes, consciousness). Information, or intellect in Maimonidian terms, is the heart of the immateriality of physical objects, whether inanimate or living. In a very Platonic sense, the addition of intellect causes consciousness to create form.

SIDEBAR: STAPP'S THEORY OF CONSCIOUSNESS BRINGS QUANTUM THEORIES TOGETHER

Both mathematicians and physicists understand the connection between the quantum world and consciousness, even if they cannot express it as such. John Von Neumann and Eugene Wigner were not reluctant to admit that in their theories, consciousness creates reality. Henry Stapp, student and associate of legendary physicists Wolfgang Pauli and Werner Heisenberg, argued that consciousness caused the collapse of quantum wave functions, and forced a choice among quantum possibilities. In other words, consciousness is the observer. Stapp proposed that consciousness is fundamental to the creation and operation of the universe.[501]

Heisenberg believed that quantum mechanics formed reality through a sequence of collapses of wave functions (quantum discontinuities). But Stapp took von Neumann's more global view that the state of the

---

[500] *Persons: Human and Divine*, p. 42

[501] *Mindful Universe: Quantum Mechanics and the Participating Observer*

universe is represented by a single wave function which is a compendium of all the wave functions that each of us can cause to collapse with our own observations. Quantum theory speaks not about matter but about our act of perceiving matter. According to Stapp there are many "knowers" in quantum theory and each act of knowledge results in a change in the state of the universe. Every individual increment in knowledge changes the universe for everyone. Everyone and everything in the universe experiences that added knowledge or information and change. This description also captures the essence of physician and philosopher Carl Jung's Collective Unconscious.

In the Heisenberg process I make a conscious choice to perform an observation. In order to know something, I am asking nature a question, and that gives me a degree of control over nature. The question that I choose to ask or do not ask lets me affect the state of the entire universe. Nature replies to my question with an observed quantity even though that answer appears to be totally random. My level of knowledge increases because I have learned something from the response. This change in the state of the universe correlates to a change in the state of my brain. There is a reduction in the wave function compatible with the information that has been learned, not just by me, but by the universe.

The foundation of Stapp's explanation of existence is consciousness. Like so many others, he sees the universe as a repository of knowledge to which we have access and upon which each consciousness exercises control.

A Single, Immaterial Consciousness that Controls Reality

Implicit in each of the many theories of consciousness is the suggestion that as we add knowledge, changing reality and the universe as we know it -- as we ask the universe questions -- we are narrowing the field of possible random replies. The implication is that growing knowledge is a key to the control of reality on an elemental level. That agrees with our intuition, Kabbalistic understanding, and the Torah's direction to its chosen people in Deuteronomy to "choose life."

This is an increasingly common conclusion among scientists. Indian physicist Amit Goswami is yet another who believes that consciousness is not a material phenomenon. Instead, consciousness collapses the quantum wave just as Von Neumann claimed. Schroedinger's famous cat could not be both dead and alive in the real world but could be in our minds, so when we make an observation we collapse the wave and see reality. He reaches this conclusion in a very logical way: If there are innumerable, discrete consciousnesses throughout creation, then how can it be determined which one of the observers of a single event collapses the wave? Or, similarly, if I tell my friend about what I saw, is it my seeing or his hearing my retelling that collapses the wave? Goswami concludes, there must be only one observer, one subject, and because of non-locality (Bell) only one "universal" consciousness in the universe. – Certainly a conclusion consistent with scripture.

Holograms also help explain this amorphous idea of the immaterial consciousness.[502] Psychologist Karl Pribram and physicist David Bohm explained that both consciousness and holograms arise from information. In the case of a hologram, that information is carried by coherent beams of light. Many think that the brain organizes information by interference patterns like those of a hologram. A hologram is a permanent record of the interference between two waves of coherent light. Each part of the hologram contains the entire image, or all of the associated information. When reilluminated with one of those coherent lights, the original three-dimensional image appears. The storage capacity of a hologram is enormous and that of consciousness seems unlimited.

David Chalmers concluded that we are going to need a new set of "psychophenomenal" laws to explain consciousness.

---

[502] David Bohm developed equations for a quantum potential acting as a kind of instantaneous pilot wave that guides particles. The wave incorporates encoded information about both local and distant events and does not fall off with distance. He believed that there is an holistic underlying implicate order whose information unfolds into the explicate order of particular fields and particles. This is like a holographic photo of which every part has three-dimensional information about the whole object (*Wholeness and the Implicate Order*, David Bohm, Boston, Routledge, 1980). Bohm's scheme shows a dramatic wholeness by allowing for nonlocal, non-causal, instantaneous connections. Events separated by space and time are unfolded from the same implicate order without direct causal connection between them.

Free Will

Perhaps the most inexplicable and exciting product of an immaterial consciousness is free will.

"Now, if you believe that the universe is not arbitrary, but is governed by definite laws, you ultimately have to combine the partial theories into a complete unified theory that will describe everything in the universe." Stephen Hawking continues, "But there is a fundamental paradox in the search for such a complete unified theory. The ideas about scientific theories . . . assume we are rational beings who are free to observe the universe as we want and to draw logical deductions from what we see. In such a scheme it is reasonable to suppose that we might progress ever closer toward the laws that govern our universe. Yet if there really is a complete unified theory, it would also presumably determine our actions. And so the theory itself would determine the outcome of our search for it! And why should it determine that we come to the right conclusions from the evidence? Might it not equally well determine that we draw the wrong conclusion? Or no conclusion at all?" [503]

If the "reductionist" is wrong, and free will does exist, then one or more of the following three approaches must be valid: First, experiential or anecdotal evidence may carry the day. But, this really isn't proof of anything. Such "evidence" is entirely subjective, and often without quantitative substance. Second, the theories of sub-atomic behaviors might be applied. Here a convincing case can be made for randomness on a quantum level that is fundamental to all existence, even when scaled up. Finally, if it can be demonstrated that consciousness is not physical, as we've been discussing above, then reductionism simply does not apply.

Medieval philosophers routinely confronted the deterministic argument and worked to explain or rationalize it. According to medieval biblical commentator and philosopher Yehuda Halevi, "All phenomena are either divine, natural, incidental or chosen."[504] "Chosen phenomena are the result of human will, occurring when one makes a decision. One's choice is included in the category of intermediary causes. Thus, one's

---

[503] A Brief History of Time, p. 12-13

[504] Kuzari, 5:19

choice has causes that trace back to the Prime Cause. And yet, this chain of causes does not force one's choice. This is because there is a free climate, and the soul is positioned between the choice and its counter-choice. It can freely choose whichever it so desires."

In 1924, Nathan Freudenthan Leopold, Jr. and Richard Albert Loeb, two wealthy students at the University of Chicago, were arrested for the kidnapping and murder of 14 year old Robert Franks. Their stated motivation was a simple desire to commit the perfect crime. The parents of the accused retained the legendary Clarence Darrow to defend them. One of the best known criminal defense lawyers in the country and an outspoken opponent of capital punishment, he argued that the defendants could not possibly have resisted the urge to murder, "because they were made that way." He also believed that criminals did not deserve to be in jail "because they cannot avoid it, on account of circumstances which are entirely beyond their control, and for which they are in no way responsible."[505]

Darrow believed that biology determined human behavior, much as does David Dennett today. Jerry Coyne of the University of Chicago states the same in his blog posts: a man cannot be held responsible for the consequences of choices that he could not otherwise have made.

In 1932, Nathaniel Cantor wrote that "man is no more responsible for becoming willful and committing a crime than the flower for becoming red and fragrant. In both cases the end products are predetermined by the nature of protoplasm and the chance of circumstances."[506]

Newtonian physics seemed to confirm the deterministic nature of life. The entire universe, from its smallest particles to its largest planets, is governed by the same fixed laws. The behavior of the smallest parts of creation determines the behavior of the whole. Change happens when the parts are rearranged while the parts themselves remain unchanged.

Modern neuroscience echoes this conclusion. Will is an illusion and the product of electrochemical activity in the brain. "Thoughts and intentions emerge from background causes of which we are unaware and over which we exert no conscious control,"[507] writes Sam Harris.

---

[505] *Darwin Day in America*

[506] *Crime: Criminals and Criminal Justice*

[507] *Free Will*

And many enlightenment philosophers and scientists (including Baron d'Holbach, Georges Buffon and Julien de la Mettrie) believed that determinism had to be true and that man was no different from other mechanisms in the universe in that regard.

Maimonides echoes Judah Halevi when he protests. "Each individual has free choice – if he wants to become righteous he can do so and if he wants to become wicked he can do so . . . many . . . think . . . that God decrees regarding each person whether he will be righteous or wicked. It is not so! Every individual can be as righteous as Moses or as wicked as Jeroboam."[508]

There are cracks in the wall of scientific determinism and some have completely abandoned the party line. "I am now convinced that theoretical physics is actual philosophy,"[509] said Nobel Prize winner Max Born.

It's a simple question then, isn't it? Do we actually make our own choices, or are they chemically and physically determined for us? Wasn't this addressed by Rabbi Akiva in the Mishna?[510] "Everything is seen (by God) and freedom of choice is given." The dissonance here is, of course, if God is all-knowing, then how can man exercise free will? Or in terms of Dr. Hawking's question, if nature is deterministic, then what freedom, if any, do we have within the system?

Piero Scaruffi,[511] author and philosopher, has a very simple and logical way to look at consciousness that is completely consistent with both the Maimonidean and Kabbalistic approaches: I am conscious. I am made of cells. Cells are made of molecules. Molecules are made of atoms, and atoms are made of elementary particles. If elementary particles are not conscious, how is it possible that many of them, assembled in molecules and cells and organs, eventually yield a conscious being like me? Sensations cannot be explained by particles. So conscious-

---

[508] Hilchot Tshuva, Rambam, Chapter 5

[509] Max Born in *Peter's Quotations*, Lawrence J. Peter, 1977

[510] Avot, 3:15; In the Babylonian Talmud, Ketuvot 30a, it is taught that "All is in the hands of heaven save for cold and heat." This refers to events that occur to man as he goes through life. "All is in the hands of heaven save for the fear of heaven," refers to "what is decreed concerning the fetus in utero, whether one will be poor or wealthy, strong or weak, foolish or wise, but not righteous or wicked." This is the interpretation of the Tosofists on TB Brachot 33b.

[511] *A Simple Theory of Consciousness*

ness must always be present. In the same way that each particle has an electrical charge, each must have a property that allows consciousness to arise.

This kind of reasoning is anathema to confirmed reductionists. Said Terry Sejnowski, Director of the Salk Institute for Biological Studies Computational Neurobiology Laboratory, "Consciousness explains things that have already been decided for you."[512] "The astonishing hypothesis is that you, your joys and your sorrows, your memories and your ambitions, your sense of personal identity and free will, are in fact no more than the behavior of a vast assembly of nerve cells and their associated molecules."[513] Daniel Dennett talks about "Fame in the Brain": There are always many potential conscious states, assemblies of neurons competing for their big moment of fame, but only one wins in this battle of survival. He says it is a mistake to think that there is any central "self" that observes or directs the show. He calls this "the Audience in the Cartesian Theater."

Steven Pinker goes further, explaining the Ghost in the Machine myth: "Neuroscience is showing that all aspects of mental life – every emotion, every thought pattern, every memory – can be tied to the physiological activity or structure of the brain. Cognitive science has shown that feats that were formerly thought to be doable by mental stuff alone can be duplicated by machines, whose motives and goals can be understood in terms of feedback and cybernetic mechanisms, and that thinking can be understood as a kind of computation. Not a computation in the way your IBM PC does computation, but computation nonetheless – a kind of fuzzy analog to parallel computation. So intelligence, which formerly seemed miraculous – something that mere matter could not possibly accomplish or explain – can now be understood as a kind of computation process."[514]

"We can't account for about half of the variation in things like personality and intellect. I suspect that this 50% of the variation that is

---

[512] *US News*, p. 61

[513] Francis Crick, 1994 book, *The Astonishing Hypothesis: The Scientific Search for the Soul*, quoted in US News, p. 61

[514] "Biology vs. the Blank Slate," *Reason Magazine*, 10/2002, Nick Gillespie & Ronald Bailey

neither in the genes nor in the family may be chance events in development, the way your brain wires itself up within the constraints of the genes." "Whether you inhaled a virus or your mother inhaled a virus, whether you got the top bunk bed or the bottom bunk bed. All kinds of uncontrollable events that may have a profound role in making us who we are."

"What we call free will is a product of particular circuits of the brain, presumably concentrated in the prefrontal lobes that respond to contingencies of responsibility and credit and blame and reward and punishment and alter their operations as a consequence."

Pinker continues drilling it home. "More to the point, you don't need to invoke a soul or some mysterious process of free will to hold people accountable. Indeed, one could argue the opposite of that: If we really are totally unconstrained – if there is a self or soul that can do what it damn well pleases – that's when holding people responsible would be futile. The soul could always choose to ignore contingencies of credit, blame, reward, or punishment."[515]

Despite this damning rationale, even Pinker's colleagues have their doubts. Says Gerald Edelman, Nobel Laureate and Director of the Neurosciences Institute in La Jolla, California, "We evolved structures that invented language but biology can take us only so far in understanding the symbol-using mind... It's not totally reductive."[516]

David Chalmers also argues "that consciousness had to be considered a fundamental category like space, time, or gravity—explicable only by special, psychophysical laws."[517] As was discussed above, at the quantum level the observer has influence, very likely through human consciousness. To reiterate the thoughts of Henry Stapp, conscious experience is an interactive event in which the observing mind has an effect on the brain.[518]

". . . the mind's resistance to simple reductive explanations lends support to the notion that it is a profoundly complex emergent system whose capacity for intentional acts and creative discoveries connects

---

[515] Pinker in *Reason Ma*gazine, "Biology and the Blank Slate"

[516] *US News, p. 61*, quoting The Conscious Mind: In Search of a Fundamental Theory

[517] Ibid.

[518] *Mind, Matter and Quantum Mechanics* per US News

it with the underlying order of reality, an order analogous . . . to the world of forms or ideas that Plato believed stood behind our shadowy and ephemeral world of appearances."[519] – Or perhaps to the quantum potential, Bohm's implicate order or the quantum vacuum – something underlying the material world, something whose energy density is estimated to be astronomical.[520]

The fact that many researchers view consciousness as tied to quantum theory's collapse of the wave function gives the laws of nature a non-algorithmic, non-formulaic element.[521] Were this not the case, humans would be little more than machines, with no consciousness at all in traditional terms – a validation of the reductionist argument.

All of us understand free will to be a function of an individual's intention. In a 1980 paper, Y. Aharonov and M. Vardi explained that "If one checks by continuous observations, if a given quantum system

---

[519] *US News*, p. 63

[520] The quantum potential carries information from the whole environment and provides direct nonlocal connections between quantum systems. It guides particles like radio waves guide a ship or plane, not by intensity but by form. It is so sensitive that the trajectories of particles seem chaotic. Bohm calls this the implicate order – like a vast sea of energy on which the explicate/physical world is just a ripple. Standard quantum theory postulates a universal quantum field (the quantum vacuum, zero-point field) underlying the material world. Its estimated energy density is astronomical. Bohm proposes a superquantum potential which imposes an organizing influence. Particles are constantly enfolding and unfolding into the implicate order and then recrystallizing. It is not measurement dependent. He rejects the positivist view that something that cannot be measured or known precisely cannot be said to exist. Our minds and consciousness do not bring quantum systems into existence. Consciousness is rooted deep in the implicate order and is present to a degree in all material forms. He thinks there may be an infinite series of implicate orders, each having both a matter aspect and a consciousness aspect. "Everything material is also mental and everything mental is also material, but there are many more infinitely subtle levels of matter than we are aware of" (R. Weber, *Dialogues with Scientists and Sages: The Search for Unity*, London, Arkana, 1990, 151). So there is no separation between matter and spirit.

[521] There seems to be a unity of information through non-locality and the implicate order. "We should expect (telepathic) phenomena to be unusual as, from the standpoint of natural selection, a person who habitually experiences other people's sensations would be less fit than a normal person. I should not be surprised if our mental insulation turned out to be a special adaptation." J.B.S. Haldane (British geneticist) in Bass, *The Mind*, p.61. Ludvik Bass was a student of Schroedinger. Clairvoyance and telepathy are called super-nonlocality by Bohm. This is much like Sheldrake's morphic resonance (R. Sheldrake, *The Presence of the Past: Morphic Resonance and the Habits of Nature*, New York, Vintage, 1989).

evolves from some initial state to some other final state along a specific trajectory . . . the result is always positive, whether or not the system would have done so on its own accord."[522] This means that vigilant observation of a quantum system prevents transition (or change) from occurring. An unstable system left on its own might be predicted to decay in microseconds, yet if watched continuously it never decays. It seems that the quantum system obeys your "intent" observation. Intent and intention are different though. Intent is focus, while intention is expectation. But, both imply the operation of will. And neuroscientist and physicist Giulio Tononi reinforces this idea by affirming the belief that consciousness is more than a combination of physical experiences, so free will must exist.

The divergence in thought among scientists is seen in the personal transformation of legendary Danish physicist, Niels Bohr. He became something of a dualist when he concluded that the quantum wave function of matter is its mental aspect. The wave of the electron is the "mind of matter." He came to believe that the duality of waves and particles is the equivalent of the duality of mind and matter.

This more global view of consciousness ties into Bohm's implicate order, Jung's collective unconscious and Rupert Sheldrake's idea of a morphogenetic field. Roger Penrose suggests that there is a set of conscious states that exists in a world of its own. Our minds have access to that world. Penrose takes a decidedly quantum "spin" on this universality however by offering that this "world of ideas" comprises quantum spin networks that encode proto-conscious states. He declares that different configurations of quantum spin geometry represent varieties of conscious experience. Access to these states -- consciousness as we know it -- originates when a self-organizing process somehow couples with neural activity, and then collapses quantum wave functions at Planck-scale geometry.

The volume of speculation and technical discussion on this topic is testimony to just how unsatisfying current expert conclusions are to the experts themselves.

---

[522] "Meaning of an Individual Feynman Path," *Physical Review Digest*, 21, no. 8, April 15, 1980, pp. 2235-2240.

Nevertheless, researchers continue to explore just how consciousness happens in a physical sense. How, in other words, does it arise from the electrical and chemical activity of neurons? What if the researchers have it backwards, and instead the activity of neurons is stimulated by something immaterial -- that which we call consciousness? What if the electrical and chemical activity that is being tracked is a response, not a cause? Kant recognized this and called the aspect of consciousness that was beyond the physical "transcendental consciousness."

"Both Susan Greenfield and I (Kristof Koch) are searching for the most appropriate neuronal correlates of consciousness. If we can find the right NCC, the direct cause and effect mechanisms that create consciousness may follow." "A coalition of pyramidal neurons linking the back and front of the cortex fires in a unique way. Different coalitions activate to represent different stimuli from the senses." Greenfield believes "Neurons across the brain fire in synchrony and prevail until a second stimulus prompts a different assembly to arise. Various assemblies coalesce and disband moment to moment, while incorporating feedback from the body."[523] The real problem that Koch and Greenfield are exploring is determining how physiological events in the brain translate into what you experience as consciousness. This is very different from finding the source of consciousness, a source which would trigger those physiological events according to one's will (see Koch's earlier remarks on consciousness).

We've seen how quantum theory offers an obvious opening for the idea of free will. The very nature of physics at sub-atomic scales is one of uncertainty so profound that it is difficult for us to grasp. However, the quantum uncertainty argument may not actually lead to a validation of free will. On the one hand, indeterminism at the quantum level may suggest free will, but that indeterminism would actually invalidate free will if the result was that our choices suddenly appear to be made in

---

[523] *Scientific American*, Oct, 2007, p. 76-83. The assemblies that Dr. Greenfield refers to here are called cell assemblies, a term coined in the 1940s by Canadian psychologist Donald Hebb. Groups of neurons that fire together tend to form cell assemblies the activities of which persist even after the event that triggered their firing is no longer present. These assemblies come to represent the triggering event. He said that the neurophysiological basis of thought was the sequential activation of various groups of cell assemblies.

a random or arbitrary manner. Free will and randomness are actually contradictory. Then too, quantum indeterminism could be caused by something on a non-physical level – our own minds or consciousness. Then free will would be the product of an active self-consciousness, or what speculators call soft determinism. Charles Laughlin[524] for example, argues that brain structures that control consciousness interact non-locally with the quantum vacuum, the great cyberspace of intellect. Others believe that there is room for free will because both consciousness and matter are born of the enormous energy potential of the quantum vacuum.

Tying Soul, Mind and Consciousness Together

There is room for free will in a quantum sense, and powerful arguments exist to support the "external" reality of consciousness and mind. Now it remains to tie the idea of the soul to this reality through the building blocks of each: information. It may be surprising to learn that the soul is just as ephemeral and perhaps imaginary to the religiously observant as it may be to the scientist or secularist.

Judaism, Christianity and Islam have wrestled for millennia with the notion of the soul. Do all humans possess a soul? Is the Jewish soul different in some way from that of the Gentile? This has, in fact, been the view of the majority of the Jewish sages since at least the Middle Ages. And this belief exists in stark contrast to the philosophical and legal focus of the Rishonim (Medieval Jewish biblical exegetes), who during their era deemphasized the external trappings of the Bible's exceptionalist/chosen mandate. "While Jewish mysticism is the source and primary expositor of this theory (the superior Jewish soul), it has achieved a ubiquitous presence not only in the writings of Kabbalists, but also in the works of thinkers found in the libraries of most observant Jews who hardly consider themselves followers of Kabbalah."[525] The *Zohar* speaks openly of this: "The children of the Holy One, blessed be

---

[524] *Journal of Scientific Exploration*, "Archetypes, Neurognosis and the Quantum Sea," C.D. Laughlin, 1996, 10:3, 375

[525] "The Soul of a Jew and the Soul of a Non-Jew, An Inconvenient Truth and the Search for an Alternative," Hanan Balk, *Hakirah, The Flatbush Journal of Jewish Law and Thought*, p. 47

He, whose souls are holy." Other nations' souls "emanate from impure sources and render impure anything that approaches them."[526]

Yehuda Halevi writes in his *Kuzari* that a convert cannot become a prophet because he does not have the same spiritual makeup as a natural born Jew. Rabbi Abraham Isaac Kook (early 20th century rabbinic leader of pre-state Israel) said, "The Jewish people are superior to all nations of the earth. . . their superiority is due to their sanctity."[527] And, Eliyau Dessler (author of *Michtav Me'Eliyahu*, or *Strive for Truth*) believed that non-Jews possess no spiritual core whatsoever, only superficiality.[528] Nevertheless, as a practical matter, this kind of thought was internalized, hardly universal, and could rarely be expressed openly in the Gentile environment in which the Jews lived and had to survive.

Contrast this with the position of Jewish rationalists led by Maimonides who believed that there was no difference between the Jewish and non-Jewish soul.[529] In his famous parable of the palace, he speaks of "every person of humankind." [530]-- not only the Jews. Maimonides comments on Amalek and the biblical requirement that the ancient tribe be wiped out: efforts must first be made to dwell with them in peace with their acceptance of the Noahide Code.[531] This belief and pattern of behavior is consistent with his displeasure with the idea of a superior Jewish soul. The acceptance of his authoritative code (*Mishne Torah*) was in part dependent on his acceptance of a rabbinic stance that rejected mystical claims, among them the belief in a unique and superior Jewish soul.

Irrespective of any difference that may exist between the Jewish and Gentile soul, science seems to be telling us that the soul, mind and consciousness have immateriality in common and may be one in the same.

Relativity theory describes matter in the context of space-time. But at the quantum level, matter interacts with a dimensionless consciousness.

---

[526] Sulam edition of the Zohar, Genesis, no. 170

[527] Midbar Shur, Jerusalem, 1997, p. 300

[528] Michtav me'Eliyahu, Sefer HaZikaron, vol.2, p.45

[529] *Maimonides' Confrontation with Mysticism*, chapter 7

[530] *Guide to the Perplexed*, 3:51

[531] Mishne Torah, Maimonides, Laws of Kings 6:1-4

If that consciousness is immaterial and is the product of the manipulation of information (just as is matter), a case for both a quantum mind and a quantum soul will have been made.[532]

What then is the soul? The mind is now so often associated with consciousness, that even among scientists the terms are often used interchangeably. The argument here is that both are immaterial, acting on the brain, and both are instrumental to choice (free will). Is the soul simply another name for the same, immaterial organ, a unique processor of information gathered in the quantum sphere that then acts on the brain and body?

If the soul (as defined by monotheistic religious tradition)is the aspect of a human that animates him and distinguishes him from other humans, animals and inanimate objects, then it is a collection of information and associated software that allows the manipulation of that information. This is consistent with the theories of Fredkin, Von Neuman, Stapp and others.

The Bible recounts that the soul was divinely "breathed" into the first man. It is this addition that differentiated Adam from earlier beings.[533] From a religious point of view, it is a supernatural aspect of otherwise natural man. Religious tradition holds that this supernatural force is part of its divine source, and is what completes the creation of the first man in the image of his Creator. The faithful believe that it retains a permanent connection to that source. As a result, the soul cannot be destroyed (nor for that matter can information).

This understanding extends to the timelessness of the soul. Once released from a living being, it continues its existence, but in another form. Let's bring the earlier discussion of mind and consciousness full circle. If the mind is the collector, repository and analyzer of information which animates life, then the soul and mind are simply terms that describe consciousness. If all are part of a single, greater consciousness, then souls can neither be created nor destroyed. They become redistributed, in whole or in part. This makes their content universally

---

[532] *The Quantum Soul: A Scientific Hypothesis*, p. 79

[533] Many religious traditions and biblical commentators believe that there were other human-like beings that preceded Adam, some of whom were contemporaries of Adam. This is, of course, consistent with what we know of evolutionary history.

available as an aspect of what must be recognized as a universal soul or consciousness. This points directly to the implicate order, quantum field, collective unconscious and several other scientific and psychological speculations.

Steven Pinker insists, "People naturally believe in the Ghost in the Machine: that we have bodies made of matter and spirits made of an ethereal something. Yes, people acknowledge that the brain is involved in mental life. But they still think of it as a pocket PC for the soul, managing information at the behest of a ghostly user. Modern neuroscience has shown that there is no user. The soul is in fact the information processing activity of the brain. New imaging techniques have tied every thought and emotion to neural activity. And any change to the brain – from strokes, drugs, electricity or surgery – will literally change your mind."[534]

Pinker cannot recognize the need to take one more step up the ladder of causality. Neural activity is triggered by something. Another physical action? This would seem to be an infinite regression. Quite to the contrary, the regression is not infinite at all. It is more likely that the brain is the hardware that carries out the software of consciousness or the soul.

"Conventional science and philosophy attempt to base consciousness strictly on classical physics, rejecting the possibility of quantum nonlocality in consciousness, including persistence outside the body as indicated by NDEs/OBEs, religious lore, and anecdotal memories suggesting reincarnation. But evidence in recent years links biological functions to quantum processes, raising the likelihood that consciousness depends on nonlocal quantum effects in the brain. That . . . suggests . . . a worldview in which consciousness or its precursors are irreducible components of reality, fundamental space-time geometry at the Planck scale."[535]

---

[534] *Newsweek*, "How to Think About the Mind," Steven Pinker, Sept. 7, 2004, p. 78,

[535] *The Quantum Soul: A scientific Hypothesis*, "Exploring Frontiers of the Mind-Brain Relationship," "Mindfulness in Behavioral Health"

Evil and Providence: Barry's Story

It had been too long. I hadn't seen my daughter in months. But now, as summer slipped away, my wife and I were looking forward to spending a few days with her, just weeks before she would deliver her first child.

The occasion was the Jewish New Year and spending it alone, now that we were "empty nesters," sounded very unappealing. So we made plans to be on the upper east side of Manhattan where we would stay with our daughter and son-in-law.

On the afternoon of the first day of Rosh Hashanah it is customary for observant Jews to symbolically toss one's sins, in the form of some spoken verses from the prophets, into a nearby body of water. Being just a short walk from the lake at New York's Central Park, we and dozens of others strolled the few city blocks from the synagogue, making new acquaintances along the way and conversing casually.

To my left walked a man in his mid-forties, laboring to drag his less responsive right side. I slowed my pace to eavesdrop on the conversation that he was having with my son-in-law. But before I could identify the subject matter, he turned to me without introduction and explained that a few years earlier, he had been the victim of a stroke. His career on Wall Street ended abruptly and he was left with both a damaged mind and body. A few years before the stroke he had undergone a spiritual metamorphosis, becoming religiously observant and devoting a year of his life to full-time religious study. "Why," he asked, "did this happen to me, and only after I became religious?" "And why hasn't God answered my prayers for recovery?"

I offered the standard set of scriptural and exegetical responses: we just cannot know the meaning of our particular misfortune and, after all, it may well be for the good; the payoff for a well-lived life in this world comes in the next world (a life after death), so the more bad that happens here, the better off one will be there; or, God does not give us situations that are too much for us to overcome.

None of these explanations is likely to be satisfying to one who is suffering right now in this world. If our misfortune actually results in good, then that good is often so obscure as to be invisible. If our reward is only to be realized in another existence beyond our own, then why must we endure suffering in this world to reach that point. If God does not give us situations which we cannot handle, then a further explanation is due the victims of crime, war and disease.

If there is a God, then it is painfully clear that He does not prevent terrible things from happening to all kinds of people, both good and bad. So my response to my new friend was weak at best, a cruel joke at worst. I left him disappointed, but no less so than was I.

"The problem of evil is something that all religions, however primitive, have to wrestle with. How did pain and misery come into the world?" asks H.L. Menken. "Why must man mourn all his days, and then pass into darkness at the end? The answer still waits, but theologians have surely made gallant efforts to formulate a credible response. They have applied many subtle devices of metaphysics, while simultaneously soothing the lowly with an endless series of facile tales about evil apples, tempting serpents and angry gods."[536]

The *Zohar*, the source book of Kabbalistic thought, responds, "You may ask: isn't a man frequently punished by God when he does not deserve it? The answer is that when a righteous man suffers it is because of the love that God has for him. He afflicts his body to strengthen his soul so that He may bring him closer in love, because it is required that the body be weak and the soul strong for a man to be loved by God. The Holy One inflicts suffering on the righteous in this world so they will merit the next world. But one who is weak of soul and strong of body is hated by God. Because He takes no pleasure in him, He inflicts no pain in this world but permits him a smooth life of ease and comfort, so that for any virtuous act, he receives his reward in this world so that none is left in the next world."[537]

Few texts venture much further than this on the issue of divine providence. Explanations of its rationale, detail and inner workings seem woefully inadequate. Believing in a good and just God is generally thought to mean believing that individuals who do good things are rewarded and that those who do bad things are punished. Bad outcomes

---

[536] *Treatise on the Gods*, p. 173

[537] Zohar, Vayeshev 180b

The Elements of Jewish Exceptionalism

for good people in this world offer the promise of an other-worldly reward. The bad who manage to succeed in this world can expect other-worldly punishment.

The Talmud explains that "Just as punishment is exacted of the wicked in the world to come, even for a slight transgression, so is it exacted of the righteous in this world, even for a slight transgression. Just as the righteous are rewarded in the next world, even for a slight mitzvah (good deed), so the wicked are rewarded in this world, even for a slight mitzva."[538]

The emphasis in this view is on the next world. But we live in *this* world, and we demand a providential explanation that makes sense to our notion of reality. It must preserve the idea of a God who is good and just, and it must distinguish between our special relationship with God as individuals and that of the community or mankind as a whole. Says Nachmanides, "For the Torah commands according to derech eretz (the way of the world) and God will secretly perform miracles for those who trust Him."[539]

Maimonides identifies a very specific formula for dictating the level of God's providential protection. God protects and oversees the life of an individual to a degree proportional to the focus of that individual on God.[540] Man's "intoxication" with God[541] guarantees the highest level of providence. But does that kind of focus assure a long, healthy life, full of success and material as well as psychic reward? Not if we believe in the purity and holiness of the greatest religious leaders, many of whom lived painfully impoverished and abbreviated lives.

For the believer, freedom of choice is necessarily connected to divine providence. He believes that God intervenes in his life and that his behavior dictates the degree and type of that intervention. Rabbi Akiva emphasized man's freedom of choice, but he left the relationship between choice and providence unresolved.

---

[538] Talmud Bavli Taanit 11a

[539] Ramban on Bereishit 12:10

[540] *Guide to the Perplexed*, 3:51

[541] As described by David Hartman in *Maimonides: Torah and the Philosophical Quest*

The Gaon of Vilna (Rabbi Eliyahu Kramer, 18th century) took a more deterministic approach toward scripture: "There is a general principle that all that was, is and will be is included in the Torah from its first word to its last. And not in generalities, but even every specific thing in creation and every individual human being -- all that will occur to him from his birth to his end, and even reincarnation!"[542]

An account in the Jerusalem Talmud records that Rabbi Shimon bar Yochai (first century, C.E.) came across a bird in a trap,[543] heard a heavenly voice call out "pardon," and said, "A bird, unless Heaven wills it, is not caught, how much more so a human being!" Both the Tanaaim (rabbis of the Mishna) and Amoraim (rabbis of the Gemara) made it clear that though God's choice was decisive, man's actions clearly influenced that choice. Furthermore, God was fully capable of confounding those choices, even though He might be saddened by it.

"A non-Jew asked Rabbi Yehoshua ben Korcha, 'Do you not agree that the Holy One blessed be He foresees what will happen?' 'Yes,' he replied. He then responded, 'But it's written, 'And it grieved him in His heart (Bereishit 6:6).' Rabbi Yehoshua said to him, 'Has a male child ever been born to you?' 'Yes,' he answered. 'And what did you do?' The non-Jew replied, 'I rejoiced and I gave a celebration for everyone to rejoice.' After which Rabbi Yehoshua said, 'But did you not realize that in the end he would die?' The non-Jew answered, 'In the hour of joy we must rejoice, in the time of mourning we must mourn.'[544]

The faithful will believe that to some degree, God influences man's choices by building on his existing pattern of behavior. The Mishna quotes Ben Azzai as saying "a mitzvah leads to a mitzva and a transgression to a transgression. For the reward of a mitzva is a mitzva and the reward of a transgression is a transgression."[545] And, "If a man sanctifies himself a little, he is sanctified much."[546] The establishment of a pattern of behavior can bring man closer to God and influence the

---

[542] *Kol Hatur*, Chapter 3

[543] Talmud Yerushalmi, Sheviit, Chapter 9

[544] Midrash Rabbah, Breishit, 27:4

[545] Mishna Avot, 4:2

[546] Talmud Bavli, Yoma, 39a

degree of providence conferred on him. Man has the free will to establish such a pattern.

The concept of free will encourages man to overrule the arbitrariness of nature and influence divine providence, a belief at odds with the deterministic argument of pure science.

Philo of Alexandria[547] saw free choice as God's gift to man for that very purpose. "After he has proved his power in subduing the evil inclination and preferring the good, and after he has realized that this power comes to him from God, God comes to his aid and helps him . . . On the other hand, the choice of evil is wholly in man's hands."[548] Biblical exegesis offers that man can choose good or evil -- it is not pre-ordained for him. "Out of the mouth of God comes neither evil nor good. Rabbi Yochanan said: From the day that the Holy One, blessed be He, said, 'see, I have set before you this day life and good,' evil and good have not come from the Holy One, blessed be He, but come of their own accord through the acts of those who do them."[549] Man is on his own, making his own bed, so to speak, while help is available from "outside the system" if he provides the right stimuli.

The last few verses of psalm 107 provide interesting insight. After a description of King David's problems and his salvation from them, the psalm concludes with a description of God's providence over nature. The performance of the Jewish people determines God's actions. He can turn the good land into a desert or the swamp into an oasis. A successful society can fail and the downtrodden can be lifted to the positions they deserve. The final line sums it up: "Let the wise man listen to these things and he will contemplate (intensely) God's kindness."[550] The wise will understand the relationship between will and providence.

It should seem obvious that a supernatural being can intervene in nature, but many religious exegetes explain that God has set up a system that minimizes or eliminates that need. Nature responds forcefully to the correct stimuli, so there is little need to either expect or depend on

---

[547] A first century CE Hellenistic biblical commentator

[548] *The Sages*, p. 253

[549] Midrash Rabbah, Devarim 34:3

[550] Psalm 107, verse 43

miracles.[551] In this construct, poor behavior strengthens the power of evil. Individual behavior has a limited impact on the strength of that evil, while collective behavior has the power to actually convert it into good.[552] Individual providence[553] can hold off evil only so long in the absence of a critical mass of collective "good" behavior.

According to this line of reasoning, God made His national, providential protection of the Jewish people conditional on the maintenance of contracts made with the biblical forefathers. "When you (Moses) die this nation will rise up and follow the alien gods of the land that they are entering. They will abandon me and violate the covenant that I have made with them."[554] As a result, "I will then show anger and abandon them. I will hide my face from them and they will be destroyed."[555] Maimonides understands that to mean that God will take away His collective providence and protection from the chosen people, leaving outcomes to chance.[556] But in the end, the people as a collective are guaranteed that God will fulfill His commitment to the covenant: "For a small moment have I forsaken you, but with great mercies will I gather you . . . says your redeemer, Hashem."[557]

MIND OVER MATTER: HOW WE ROUTINELY MANIPULATE THE INFORMATION THAT DETERMINES OUTCOMES

The Japanese monkey, Macaca Fuscata, has been observed in the wild for decades.

"In 1952, on the island of Koshima, scientists were providing monkeys with sweet potatoes dropped in the sand. The monkeys liked the

---

[551] Ramban on Bamidbar 26:11. Rabbi Joseph Dov Soloveitchik spoke in his *The Emergence of Ethical Man* of the wholly natural origin of what we think of as miracles – only timing and circumstance cause us to regard them as miraculous.

[552] See Ramchal's *Daat Tevunot* for a book-length presentation of this principle.

[553] "That God makes nature serve those who believe in Him is a major Torah principle and a basis of our faith in God's intervening in individual destiny." Rabbi Yoseph Albo, Sefer Ikkarim, IV:22

[554] Devarim, 31:16

[555] Devarim, 31:17

[556] *Guide to the Perplexed*, 1:23, 3:51

[557] Isaiah, 54:7

The Elements of Jewish Exceptionalism

taste of the raw sweet potatoes, but they found the dirt unpleasant. An 18 year-old female named Imo found she could solve the problem by washing the potatoes in a nearby stream. She taught this trick to her mother. Her playmates also learned this new way and they taught their mothers, too . . . Between 1952 and 1958, all the young monkeys (but not all parents) learned to wash the sandy sweet potatoes to make them more palatable . . . Let's suppose that (in the Summer of 1958) when the sun rose one morning there were 99 monkeys on Koshima Island who had learned to wash their sweet potatoes . . . later that morning the hundredth monkey learned to wash potatoes. Then it happened! By that evening almost everyone in the tribe was washing sweet potatoes before eating them. The added energy of this hundredth monkey somehow created an ideological breakthrough!

The most surprising thing observed by these scientists was that the habit of washing sweet potatoes then spontaneously jumped over the sea -- colonies of monkeys on other islands and the mainland . . . began washing their sweet potatoes."[558]

Ken Keyes retells this story in order to rally society to fight the proliferation of nuclear weapons. He promotes the influence of a critical mass of popular thought, connecting with an amorphous collective intelligence in the mold of Carl Jung's collective unconscious. For the good of mankind, we want to believe that the Hundredth Monkey Phenomenon is fact. Unfortunately, it appears that some of the details of the original research were misrepresented, and that there is no empirical data to validate the spontaneous transmission of behavior either within the original colony or to other unrelated colonies of monkeys.[559]

But to many people, this phenomenon is far too appealing to reject simply because it has no empirical support! There seems to be an irresistible intuitive aura of truth about it. Just as has been discussed above as an outgrowth of the quantum aspects of nature, many authors, artists and psychologists are convinced that concentrated collective thought can influence nature -- that lines of communication are open and active on every level of creation at all times. To others, such ideas sound like

---

[558] *The Hundredth Monkey*, pp.11-16, taken from a report by Lyall Watson in *Lifetide*, Bantam Books, 1980, pp. 147-8.

[559] "The Hundredth Monkey Revisited," NHNE, 4/11/97, James Gregory

something better found in a science fiction novel than in a scientific research paper or a serious religious commentary. But, the antiquity of the religious incarnation of this theory is well documented.

"Thought, when intensely concentrated can exert great influence. Every faculty of the mind, down to the inner most point must be focused without distraction. When many people do this, their thoughts can actually force something to happen."[560]

Clearly, there are all kinds of natural phenomena of which we are only dimly aware. For example, empirical data show that laboratory rats take some time to solve a new maze, but their successors, having never seen the new maze, solve it with increasing ease. It seems that some kind of communication -- independent of contact and distance -- must be present.

There must be a reservoir of thought and ideas that each of us can tap into. If not, how can so many weird, natural phenomena be explained? We know that only a small number of "organic crystals" occur naturally. Others are synthesized in laboratories. Of those that are artificially synthesized, it is observed that crystals often take hours to form from their original "mother liquors." But when this process is subsequently repeated under identical circumstances elsewhere, the time required for crystallization rapidly falls until these exact crystal patterns are being routinely produced around the world within a few minutes of preparation. How can this be explained?

In 1978 English botanist Rupert Sheldrake advanced the notion of the Morphogenetic Field, a previously untheorized energy field. Through a process that he called Formative Causation this field influences living organisms by being a source for the information necessary to shape the exact form of a living thing, and influence its interaction with other beings. The theory has attracted as much ridicule as support. Some may see in this theory commonality with the theory of the collective unconscious. Unlike Freud, his student Carl Jung believed in a collective psychology, something that he called "a reservoir of the experiences of our species." He believed that part of a person's unconscious was common to all humans, containing forms and symbols common to all cultures.

---

[560] *Rabbi Nachman's Stories*, p. 170

Both Sheldrake's and Jung's theories may simply be interpretations of what physics now calls the quantum field or Bohm's implicate order.

It appears that this collective, "Hundredth Monkey" process is mimicked at the inorganic level as well. When an atom of the radioactive isotope Uranium 235 decays, it produces the nuclei of two lighter elements along with two neutrons. These escaping neutrons can induce fission of other U235 nuclei. The process repeats and spreads in a chain reaction. On the larger scale, this results in a "critical mass" of U235 in the form of a solid sphere 7.25 inches in diameter. Anything smaller will get hot without exploding, while anything larger results in a chain reaction so rapid that it becomes an atomic bomb.

In each of the cases mentioned here the density or concentration of energy triggers a reaction -- sometimes mechanical, sometimes biological and sometimes cognitive. So, can the mind be used to deliberately alter creation? Can various aspects of society be "mentally" engineered?

The placebo effect is a euphemism for the influence of that concentration of energy, or thought on the physical. Frequently demonstrated in studies of pain relieving or therapeutic drugs, it is just as easily displayed in other human behaviors.

Harvard psychologist Ellen Langer surveyed 84 hotel maids and found that 67 percent of them reported that they did not exercise with any regularity. Most claimed to get no exercise at all. Yet all of the women in her study exceeded the U.S. Surgeon General's recommendation for daily exercise by virtue of their normal daily work routines. Nevertheless, despite that work load, the women showed no discernable physical benefit from that activity. In fact, physical measurements taken by Langer's team corresponded to the level of exercise perceived by the maids rather than the actual level experienced. Langer then divided the maids into two equal groups, one of which was supplied with information on the actual level of exercise they were experiencing and its expected benefit. The other group was not provided any such information. When Langer's team returned a month later to collect new data, they found that the first group had achieved dramatic, measurable improvements in their physical condition, without changing anything about their work routines, diet or exercise. Langer believes that their altered mindsets altered their physical reality.[561]

A Hebrew University Study demonstrated that stress hormone levels in the bodies of marching soldiers always reflected the distances they were told they had walked, not the actual distances covered. A NASA study in anticipation of the 1980 Winter Olympics demonstrated that Russian athletes who used more mental imagery techniques in training outperformed their traditionally trained teammates. Since that time, similar results have been documented for basketball players and golfers, among others.

Psychologist Bruno Klopfer's report on his patient (Wright) who suffered from advanced cancer of the lymph nodes is one of many examples of both the placebo effect and the mind's inability to distinguish between an imagined reality and an actual one. Wright was told about a new experimental drug, begged to have it, and immediately after its administration, began a dramatic recovery. His tumors disappeared within days. Wright remained completely well until articles on the ineffectiveness of his new wonder drug began appearing. He became depressed and soon relapsed. Dr. Klopfer then told Wright that the reported ineffectiveness was due to an adulterated batch of the drug and administered a pure batch (which was actually just water). Wright again recovered completely and remained healthy until hearing that the AMA debunked the drug's effectiveness. He died two days later.

In 1962, Vittorio Michelli's cancerous hip tumor was so large and destructive that he was sent home from a hospital without treatment. However, after several dippings in the spring at Lourdes he began a recovery that saw the tumor disappear and the hip bone reconstruct itself.

On thousands of occasions Christians have evidenced the supposed wounds of Christ's crucifixion on their own bodies (these markings are called stigmata). All those who have such wounds have them on their hands and feet where tradition and art have taught that Christ was nailed on the cross. Some even display nail-like protuberances from the wounds. In fact, Roman custom, supported by skeletal remains demonstrates that nails were placed through the wrists, not the hands.

---

[561] NPR, Your Health, 1/3/2008, Morning Edition, Alix Spiegel, "Hotel Maids Challenge the Placebo Effect"

Anecdote after anecdote confirms that a person's belief can dictate health, death or recovery and can physically change the human body. In a further example, to this day The Phuket Vegetarian Festival (Thailand) is observed with its massive self-mutilations by participants (with only stiches or less to show for it).

The Power of Group Think

"... nearly all of the miracles performed by saints and wonder-workers of the world's great religions have also been duplicated by psychics. This suggests that, as with stigmata, miracles are produced by forces lying deep in the human mind, forces that are latent in all of us."[562] William McDougall, pioneer in instinct and social psychology in the early 20th century, attributed the accomplishment of religious miracles to collective psychic powers.[563] And consistent with our discussion of the pervasiveness of consciousness, he believed that there is a mental aspect to all matter.

In the 1960s and '70s forms of far-Eastern meditation were popularized in the West in response to increasing social, political, economic and generational pressures. The Beatles made the Maharishi Mahesh Yogi a pop icon, and enabled his movement to attract followers and sufficient financial support to open three university campuses. Like any self-respecting academic institution, Maharishi International University conducts academic research. Its researchers hypothesized and then tested a phenomenon somewhat cynically called the "Maharishi Effect." They took the Critical Density Effect (examined above) and elevated it to the collective human theater.

In 1960, the Maharishi began suggesting that if 1% of any community regularly practiced a particular kind of meditation, the result would be a measurable decline in street and domestic violence, car accidents, burglaries, hospital admissions and the like. To test this theory, groups of 100 or more of the Maharishi's faithful were sent to meditate together in towns of under 5000 inhabitants all over the United States. In anticipation of their arrival, local government officials collected relevant

---

[562] *The Holographic Universe*, p. 120

[563] *The Group Mind*

statistical data that could be compared to data compiled during and after the experiment. Positive results have been demonstrable, consistent and compiled in quasi-scientific papers. The test communities have regularly requested repeat visits.[564]

The "Maharishi Effect" seems to demonstrate what observant Jews have always believed, and what prayer congregations of Christians and Muslims promote. Both the power and intent inherent in Judaism, for example, are in part products of community. Scriptural and traditional directives are to be practiced as a group. Described in scripture as a single entity, the Jewish people can only achieve its goals with the co-operation of its parts. Jews are described biblically as an *am* (nation), *Bnei Yisrael* (Children of Israel), *eidah* (congregation), *klal* (whole, totality), *agudah* (association, tied together), *knesset* (gathering), *kahal* (public assembly), *kehilla* (community) and a number of other collective Hebrew terms.

The commandments of the first five books of the Hebrew Bible are aimed at common well-being and a relationship between the community and its creator. Both scripture and tradition testify to the strength of numbers. A critical mass of Jews with a common goal influences the outcome of prayer and the efficacy of action. One cannot compare the effectiveness of a small group in the fulfillment of a scriptural commandment to that of a large group.[565] When the prescriptions of scripture are carried out correctly and collectively, "five will chase 100, and 100 will chase 10,000."[566] The more who are united in thought and action, the more effective the thoughts and actions.[567]

Jews stand three times a day and pray to their Creator. Each prayer is expressed in the plural. Each request is a request for the Jewish collective. "Prayer is a very awesome and profound thing . . . The future redemption, may it come speedily and in our days is also dependent upon (it). Therefore it is most correct to make every effort and be careful

---

[564] Descriptions of these controlled studies have appeared in *The Journal of Conflict Resolution*, *Journal of Crime and Justice*, *Social Indicators Research* and the *Journal of Mind and Behavior*.

[565] Torat Kohanim on Vayikra 26:8

[566] Vayikra 26:8

[567] Rashi on Vayikra 26:8

to pray in a *Beit Knesset* (synagogue) . . . for here dwells the *Shekinah* (God's presence). There a man will be ten times more prosperous in the recitation of his prayers than he would be if he prayed at home. In the *Beit Knesset* there are two benefits, the holiness of the place as well as 'In the multitudes of the people is the glory of the King.'"[568]

The Talmud too repeatedly testifies to the power of sincere and concentrated collective spiritual action. "On the day that Rebbe (Rabbi Yehuda HaNasi, redactor of the Mishna) died, the rabbis declared a public fast and offered prayers for heavenly mercy (to preclude his death) . . . Rebbe's handmaid went to the roof and prayed . . . but when she saw how painful taking off and putting on his tefillin was (because he needed to attend to his bodily functions so frequently) . . . she took up a jar and threw it down from the roof to the ground. For a moment they (the other rabbis) stopped praying and the soul of Rebbe departed."[569]

In the prophecy of Ezekiel (37:15-19) it is predicted that two sticks representing the divided people of Israel (the Northern Kingdom that split off after the death of Solomon) and Judah[570] will be brought together and the Jews will be reunited eternally. The Midrash tells us that if the Jews become a single unit, we should prepare for the final redemption.[571]

Jewish history is one of fragmentation and polarization -- not just politically and ideologically, but socially. Yet, the language of creation is the language of unity and hope, as is the language of Kabbalah. The *Kav*, that primordial line or ray of supernal light that activated all creation, is the root of the Hebrew word *tikva*, or hope. All Jews are bound together by this ray of hope,[572] a ray that also binds creation to the future of mankind.

---

[568] Ben Ish Chai, Introduction to Parshat Miketz

[569] Talmud Bavli, Ketubot 104a

[570] After the death of King Solomon, the Jewish Commonwealth split into two Kingdoms: Israel in the north and Judah, with the Temple and its capital of Jerusalem in the South. The ten tribes of the Northern Kingdom were eventually conquered and exiled by the Assyrians.

[571] Midrash, Bereishit Rabbah 98:2

[572] *Orot*, as noted by Rav Yitzchak Hutner, cited in note 304, p. 269

LINKING FREE WILL, PROVIDENCE AND THE PROBLEM OF EVIL

The classic monotheist believes in a God of perfection -- all-knowing, all-powerful and completely good. Yet, how can a perfect God permit or be responsible for the imperfections that characterize creation? What place does the existence of evil have in the purview of a providential God? Why is suffering allowed in what Maimonides and Leibniz believed to be the best of all possible worlds? Death alone negates all goodness, beauty and accomplishment. The presence of imperfection or evil in the world seems to preclude the existence of the God of Abraham, Jesus and Mohammed.

Theists argue that true good cannot be achieved without evil in the world, and that as imperfect beings we simply cannot comprehend this divine fact. Yet rational man surely rejects the idea of a greater good emerging from the Holocaust or the sufferings of disease or war.

Some argue that if God is omnipotent he could have created people possessed of free will but without sin.[573] The true freedom permitted to man by God allows for the unfettered exercise of that freedom. Otherwise, God would be manipulating our choices in order to eliminate evil or amoral or immoral outcomes.[574] And if God regularly interferes with our pursuit of our own objectives, then His objectives cannot be met. Critics will argue that God is relinquishing sovereignty or omniscience if by permitting this freedom, He does not have knowledge of our choices until after they are made. These arguments minimize God's perfection, and simultaneously absolve Him of responsibility for evil.

How then can God's omniscience and sovereignty be preserved in the face of free will and evil?

The answer is a by-product of our research on the sub-atomic world. If we extrapolate from arguments espousing the immateriality of the sub-atomic realm and the "illusion" of time that was discussed earlier,[575]

---

[573] *Mind Magazine*, Volume 64 (254), p. 209, J. L. Mackie

[574] *God, Freedom and Evil*

[575] "The gap between the scientific understanding of time and our everyday understanding of time has troubled thinkers throughout history. It has widened as physicists have gradually stripped time of most of the attributes we commonly ascribe to it. Now the rift between the time of physics and the time of experience is reaching its logical

we can then begin to form a conception of a Supreme Being who stands outside of time,[576] indifferent to our distinctions of past, present and future.[577] He also equips every aspect of creation with consciousness and free will so that anything that is logically possible can occur, with only the probabilities of such occurrences remaining a mystery. [578]

According to this line of reasoning, man lives under a "bell curve" of experience, action and probability. He and nature have been assigned a model of preferred individual and communal behavior. Conforming to that behavioral model moves an individual many standard deviations from the mean, toward the very outermost reaches of the curve. There, the probabilities of an undesirable outcome are reduced. But this particular response to pure free will does not eliminate the possibility of evil entirely, it just reduces its probability. Probability, therefore, is a product of behavior and adherence to the model. One's position "relative to the mean" allows for the clash of myriad free wills and the occasional causalities that may result.

---

conclusion, for many in theoretical physics have come to believe that time fundamentally does not even exist." Craig Callender, "Is Time an Illusion?" *Scientific American*, June, 2010, p. 59. Physicists have also linked time to the second law of thermodynamics -- as the universe gets older, it becomes more disordered. But physicists cannot explain why order is found in the past and disorder in the future; why the universe began hot and homogeneous and is cool and lumpy today; why we remember the past and not the future or why we cannot turn an omelet into an egg.

[576] "The quantum world consists of vibrations that, like musical notes, are nothing at an instant and require time in order to exist. It is a world of dynamic flux in which particles come and go. It is a world of probability states; only the passage of time will disclose which of the alternative potentialities will be actualized. Time is not the unwinding of a predetermined scroll of events but the novel coming-to-be of unpredictable events in history." *The End of Time*, p. 192

[577] Boethius, Consolation, Bk. V, pr. 6; God exists entirely outside of time in an eternal present to which all that occurs in time is before Him. All of history occurs to Him as if at the present.

[578] Fred-Alan Wolf believes that additional information was needed to start life, not just evolution and natural selection. That information came from the future. Wolf says nature produced the right organisms to survive in their environment because information flowed back from the future to the present about which organisms made sense. When we "fix" an outcome today of an observation we are forcing a particular past on that object. Our conscious mind can create a past from all the possible pasts.

A Supreme Being who views time as undifferentiated sees choices instantaneously reflected in a "timeline,"[579] that displays all associated outcomes. The "immateriality" of creation allows man to manipulate his own reality (on his own primitive level) to better reflect his understanding of God's model. Pharmaceutical efficacy, stigmata, and extra-sensory or extra-physical experiences all reflect the beginnings of man's divine manipulation of his reality -- the "image" of God in which he was created.

The timeless existence of consciousness and the mind, and its participation in a collective unconscious, implicate order, morphogenetic field or quantum vacuum allows for the existence of providence. The consciousness/mind/soul of man provides providence, both on an individual and communal level, by manipulating the immaterial essence of all creation as provided by God. God has created a system that does not require His direct intervention. It requires our knowledge of His expectations and the substance of His creation. The first was supplied by revelation. The latter is being uncovered by mathematics and its interpretation.[580] In the context of

---

[579] As noted above, time may not exist at a fundamental, physical level at all. Instead, just as a solid object is primarily empty space and its solidity is an emergent property of its particles, so too time may be an emergent property of the basic building blocks of the world – here argued to be information. The basic laws of the universe seem to be unaffected by the direction of time. That is, from the point of view of physics, the past, present and future exist simultaneously. So a divine timeline, a timeline observed by the Creator of the natural laws, would not reflect time at all.

[580] The quantum vacuum provides a home for the various "elementary particles" composed of quarks that seem to be temporary manifestations of shifting patterns of waves that combine at one point, dissolve again, and recombine elsewhere. A particle begins to look like a local outcropping of a continuous substratum of vibratory energy... as more complex systems are built up, new properties appear that were not foreshadowed in the parts alone. (The End of Time, p.174) Particles and fields exist (Bohm). Particles have a complex inner structure and are always accompanied by a quantum wave field. They are acted upon by classical electromagnetic forces and also by the quantum potential determined by their quantum field (this obeys Schroedinger's equation). The quantum potential carries information from the whole environment and provides direct nonlocal connections between quantum systems. The quantum potential is sensitive and complex so particle trajectories appear chaotic. This is what Bohm calls the Implicate Order (a vast ocean of energy on which the physical or explicate world is just a ripple). Standard quantum theory postulates a universal quantum field (the quantum vacuum or zero point field) underlying the material world. Very little is known about the quantum vacuum but its energy density is estimated to be an astronomical 10 to 10^{8th} J/cm3.

the Pentateuch, being created in His image implies a "love of God," prayer, meditation and other methodologies that acknowledge a Creator, blur the boundaries between the material and immaterial, and provoke and create providential responses.

The approach to providence explained by Maimonides in his Guide to the Perplexed, speaks to the nature of creation described above and to the ability of man to draw providence from his interaction with it.

The two major religious approaches to providence are represented by Maimonides and Nachmanides. Ramban (Nachmanides) believed that miracles were interruptions in the natural order, while Maimonides saw them couched in the cause and effect of the natural order. The Ramban ties his understanding to reward and punishment: "A person has no portion in the Torah of Moses unless he believes that all our matters and circumstances are miracles and they do not follow nature or the general custom of the world . . . rather, if one does mitzvot he will succeed due to the reward he merits."[581]

To Maimonides, evil constitutes the lack of goodness and perfection. He looks at the "whole" of the world and sees the wisdom that allows evil to contribute to it. [582] Maimonides believed that in our realm only humans are subject to God's providence. Other species are preserved, but everything else is left to chance (*keri*, in Hebrew, as was described above). "God has put into place a system that is there for individual human beings to take advantage of or not, as they choose. And it is the virtuous – understood as those who pursue intellectual virtue, and not merely moral virtue – who choose to do so, while all others are left without its protection."[583]

Individual providence according to Maimonides is derived from the emanation and overflow of knowledge from God. Receiving this overflow of intellect provides providential protection. The more man turns to God and attends to this overflow of knowledge, the more providential will be his fate. ". . . Divine providence does not watch in an equal manner over all the individuals of the human species, but providence

---

[581] Nachmanides commentary on Exodus, 13:16

[582] *Guide to the Perplexed*, 3:12

[583] *The Order of Nature and Moral Luck: Maimonides on Divine Providence*, p. 9

is graded as their human perfection is graded."[584] ". . . when one is not attending to God (either because one has never made the effort or because, having achieved the connection, one has temporarily become distracted, perhaps by the pleasures of the senses), one is abandoned to chance and left to one's own devices in the face of the slings and arrows of outrageous fortune. The person who is not experiencing the overflow is not enjoying its benefits. He is at the mercy of nature's elements and his well-being is subject to whatever may or may not come his way."[585]

To the degree he ignores the intellectual overflow, he positions himself nearer the mean, under the most expansive part of the bell curve. People who do not focus on that divine, intellectual overflow enjoy only the level of providence that ensures the survival of their species. They have left themselves completely subject to chance. Maimonides says that the individual who is truly connected to the overflow can make himself immune to the random "evil" occurrences in life.[586]

Professor Steven Nadler argues on behalf of Maimonides that the perfect intellect for which man strives attains theoretical knowledge of both natural and divine science. This includes knowledge of the cosmos and the order of things in our world – the workings of nature. "A person with a deep knowledge of nature will have extraordinarily accurate predictive power, and thus will know what the course of nature typically brings in certain circumstances. He will rarely be taken by surprise, and thus in the worldly conditions of his life moral luck will be reduced to an absolute minimum."[587] Others have interpreted these passages to mean that either astrological interaction or other extra-natural intimations will direct the virtuous man. Instead, Nadler concludes that providence comes from "a kind of intellectual reasoning about the order of nature, a reasoning grounded in an understanding of the principles of the cosmos and leading to a predictive and practical conclusion."[588] Maimonides,

---

[584] *Guide to the Perplexed*, 3:18

[585] *The Order of Nature and Moral Luck: Maimonides on Divine Providence*,, p. 11

[586] *Guide to the Perplexed*, 3:51; This claim was the subject of a letter from Samuel Ibn Tibbon, translator of the Guide from Arabic to Hebrew, to Maimonides in 1199.

[587] *The Order of Nature and Moral Luck: Maimonides on Divine Providence*, p. 17

[588] Ibid, p. 18 on The Guide 3:51

then, believes that a person who suffers from evil deserves it – his actions have taken him away from the protection of providence.

4. THE MESSIANIC IDEA

Messianism is religion's answer to oppression, persecution and desperation. It is common to the idea of the Christian savior and to the Muslim 12th Calif. The idea that the salvation of an entire people could be led by a human agent of God is particularly Jewish. Moses models that role in the Bible, and the prophets herald a future redemption along just those lines. Midrash and mysticism offer prescriptions and predictions of a redemption initiated by a key figure. This Messiah, a savior-king, is a legacy of Torah references to God's perpetual attention to the suffering of His chosen people.

The apocryphal book of Enoch (1:48) records the words of God at the time of redemption: "I shall deliver them (kings of the other nations) into the hands of my elect ones like grass in the fire . . . and they shall not rise up again . . . for they have denied the Lord of the Spirits and his Messiah." Chapter 7 of the book of Daniel says of the Messiah, "And he came to the Ancient One and was presented before him. To him was given dominion and glory and kingship that all peoples, nations and languages should serve him."

The Book of Samuel (II Samuel 7; 23:1-3, 5) declares that God had specifically chosen David and his line to reign over the Jews forever and over the other nations of the world as well (II Samuel 22; 44-51 and Psalm 18 and Psalm 2). David and his offspring are identified as "anointed" in the same fashion as are foreign kings. The prophets Amos, Isaiah, Hosea, Jeremiah and Ezekiel refer to the divine nature of the perpetual reign of the Davidic dynasty from which the Messiah will rise. One of the 19 blessings of the Amidah prayer, said three times daily by observant Jews, requests the arrival of the Messiah from the line of David and the redemption that will then be introduced.

In chapter 9 of the prophecy of Isaiah the messianic idea evolves, shifting its focus from the lineage of the monarch to his qualities. This focus grows in post-biblical writings with numerous references to the salvation of the Jews in the absence of an individual who has been anointed to lead that process. The book of Zechariah refers to both a high priest and a messianic king -- the traditional scriptural leadership pairing. At the dawn of the Christian era, the Dead Sea sect carried this tradition forward, echoing an unnamed prophet of the "last days." The usurping of the monarchy by the priestly Hasmonean family a century and a half before the Common Era is very likely the reason for the later restoration and promulgation of references to the Davidic Messiah.

If the messianic process takes place in the absence of supernatural intervention (as Maimonides believed) then information is the critical tool directing the principals, the process and the outcome. The guardians of Jewish tradition expect the Messiah to be a human. Events that precede his arrival and the actions he takes thereafter are expected to be wholly natural. Our earlier discussion of the science and Kabbalah of exceptionalism anticipates a messianic figure who has access to all information necessary for the recognition and completion of his task – information that already exists in the implicate order/collective unconscious. He simply needs to become aware of his role and then access the information that is germane.

Historically, there have been many candidates for the job. Jewish tradition tells of several (Moses, Hezekiah, Bar Kochba), holds the belief that there is a potential Messiah in every generation, and gives reasons for the failures of some who did not successfully assume that role. There have been many false Messiahs as well who themselves believed in their mission but disappointed their followers. Traditions that require their messianic candidates to come back to life after dying or to be supernaturally delivered to mankind do not meet the basic Jewish criteria.

Maimonides --who in the 12th century included the messianic idea as one of his 13 principles of Jewish faith -- noted that it is alluded to twice in the Torah -- in the story of the evil prophet Bilaam[589] and near the end of the book of Deuteronomy.[590] In addition, the books of the prophets are filled with references to the subject. The Gemara points out that "There is no difference between this world and the Messianic Age, except with regard to our subjugation by other governments."[591] The Messiah will be

---

[589] Bamidbar, 24:7

[590] Devarim, 30:3

[591] Talmud Bavli, Berachot, 34b

from the line of King David, and with the help of God will defeat enemy nations and solidly reestablish Israel as a Jewish state adhering to the laws of the Torah. The other nations of the world will admire the Jews and honor God at his rebuilt Temple in Jerusalem. The world's general knowledge of God, His actions and proximity, will grow dramatically, and there will be no further reason for wars among nations.

The commentary of Maimonides points to Deuteronomy 30:5 as a summary statement that includes all of the prophetic predictions relevant to that time: "God will restore your fortunes, have mercy on you, and gather you (from where you've been scattered). If He banished you to the ends of the heavens (He will still gather you and bring you back). The Lord your God will bring you (back to the land of Israel)." In Numbers 24:17, there seems to be a prophecy about King David and a later descendant who will ultimately redeem the Jews: "A star shall come forth from Jacob, a scepter shall arise from Israel. He shall smite the legions of Moav."

According to most opinions then, the Messianic Era will be a time of peace between nations in a non-miraculous, natural environment. This age will begin with the war of Gog and Magog, as described in the book of Ezekiel, during which the nations of the world will attack Jerusalem.[592] Magog was one of the grandsons of Noah. One of his descendants, Gog, will become a king and will assemble a coalition of nations to attack Israel and Jerusalem. According to one of the later Midrashim[593] unrest in the Islamic world will push the assembled nations to war. While they will begin by battling with the Jews, the nations will ultimately find themselves battling against God, who will come to the aid of His chosen nation. The prophetic outcome of that war will result in 7 months of burying the enemy's dead.[594] According to various other prophecies, the return of the prophet Elijah (who tradition has it never actually died) will either precede or follow this war. Maimonides

---

[592] The Chofetz Chaim, Rabbi Israel Meir Kagan, who was a leader of the early twentieth century European Jewish community, believed that the war of Gog and Magog would actually be a series of three wars; the first of which he believed was World War I. Though he died in 1933, he predicted that the second of these wars would begin in the late 1930's and would be much more destructive than the first.

[593] Yalkut Shimoni, Yeshayahu; 60:499

[594] Ezekiel 39:12

(in his Letter to Yemen) believed that the Messiah would arrive on the scene only after the conclusion of this war. The details are obscure.

Maimonides concluded, "In all cases such as these, no man knows what will happen until the time comes. These things were purposely left unclear by the prophets. Our sages (the rabbis of the Mishna and Gemara) did not have a clear tradition in this subject either, and could only come to some conclusion by derivation from many biblical sources. This is why there are so many opinions on these matters." [595]

Jewish tradition offers two alternatives for the ultimate redemption. Both are drawn from a single verse from the prophets: "I, God, will hasten it (the redemption) in its time."[596] To "hasten" it seems to mean that God can be prompted to move up His timetable of redemption. But, "in its time" implies just the opposite, that God's original plan will remain intact, with the redemption coming at a preset time, irrespective of circumstances. The conclusion drawn by most biblical commentators is that God has, in fact, predetermined a time for redemption. And for a variety of reasons, that is thought to be sometime before the year 6000 (2240 on the Gregorian calendar). This date and the history that has brought us close to it is the source of volumes of Kabbalistic interpretation. However, say the commentators, God has offered the Jewish people an opportunity to actually reduce the time until redemption. To activate that alternative His chosen people must follow His Mosaic prescription. Only then will circumstances be conducive to an early, complete and final redemption.

Given these two alternatives, the Bible and its commentators follow dual approaches whose pronouncements often seem contradictory.

Maimonides and others are convinced that the Messianic Era will commence in a natural manner: "Behold, your king comes to you . . . in humility and riding on a donkey."[597] If the Jews continue to ignore their biblical destiny, redemption will come, but through a seemingly ordinary progression of natural, historical events. And history speaks to a succession of such events that have already taken place (the Holocaust, the ingathering of the Jews and the creation of the State of Israel). On

---

[595] Mishne Torah, Hilchot Melachim, 12:2

[596] Isaiah 60:22

[597] Zecharia 9:9

the other hand, there are those who assume that the coming of the Messiah will be accompanied by miracles and wonders: In the Book of Daniel we find, "In the night visions, I beheld the likeness of a human coming with clouds from heaven."[598] The understanding in this case is that should the Jewish people overwhelmingly return to a Torah way of life then the Messiah will arrive in a miraculous way.

The Vilna Gaon (Eliyahu Kramer, 1720-1797) understood himself to be living in the time immediately preceding the Messianic Era. As such, he drew on the words of others before him and developed both an analysis of history and a program to hasten the redemption. The "Gra" (Gaon Rabbi Eliyahu, the Vilna Gaon) drew on the Zohar and Kabbalistic thought to compare the seven days of creation to the 7,000 years (the classical Kabbalistic notion) of this world's existence (see the earlier kabbalistic/scientific discussion of creation). Mankind has six days or 6,000 years to play out the divine plan before the arrival of the Sabbath, corresponding to the seventh millennium -- 1000 years of a new existence. The Zohar speaks of the year 5500 (1740) as being the beginning of a period that it calls the "footsteps of the Messiah." The analysis begins with the correspondence of each day -- 24 hours -- to 1000 years of the world's existence. At 6 am (half way through a 6 pm to 6 pm, 24 hour day) on the eve of the Sabbath (1740 or twenty years after his birth) the Gra claims to have been visited for the first time by a maggid (a heavenly messenger) who began to teach him the secrets of the Torah.

Some of the prerequisites that he and others saw for the messianic redemption are already in place, but the Gra believed almost 300 years ago that the Messianic Era was close. What signs of redemption are recognizable today?

"The eyes of the blind will be opened and the ears of the deaf unplugged. The crippled man will jump like the deer and the tongue of the speechless will sing."[599] Does this refer to technological developments in health care? The Zohar predicted that in the year 5600 (1840) a heavenly influence would be introduced by God into the world. If worthy, the Jews would use this new knowledge or power to step up messianic

---

[598] Daniel 7:13

[599] Chabakkuk 35:5,6

progress. If not, the Gentile nations would husband this divine influence for their own purposes. This date correlates to the generally accepted commencement of the Industrial Revolution.

"In our day there take place in the course of a short period innovations which in the past would have required centuries. We see that the wheel of time revolves now with lightning speed. In Heaven there were set up from the beginning until now innumerable accounts. Before the Messiah can come, all these accounts must be paid, because the redemption will annul the evil inclination and, consequently, all the affairs of this world which are subordinate to the spiritual strife within man, will cease."[600]

The return of the Jews to the Land of Israel, the rehabilitation of the land and the assistance given by other nations in that effort at the advent of the Messianic Era are all foretold in the prophets.[601]

It is also said that these pre-messianic times will be characterized by a general decline in morality.[602] Children will show little respect to their parents and will scorn them.[603] Even the re-established state of Israel will abandon the Torah.[604] But God is merciful and He will still send the Messiah -- either because Jews have repented voluntarily[605] or because an oppressive ruler causes them to do so.[606]

We really don't have very much to go on. Those descriptions that we do have will only be verifiable in retrospect. Jewish tradition offers that the Messiah will come from the line of King David, from the tribe of Judah.[607] He will not be a miracle worker and he will not "bring about new phenomena within the world, resurrect the dead, or perform other similar deeds," said Maimonides.[608]

---

[600] Elchanan Wasserman quoting the Chofetz Chaim in Ikvessa D'Meshicha

[601] Ezekiel, 37:21 and 16:55

[602] Amos, 8:11

[603] Mishne, Sota, 9:15

[604] Mishne, Sota, 9:15

[605] Mishne Torah, Hilchot Tshuva, 7:5

[606] Talmud Bavli Sanhedrin, 97b

[607] Chronicles 22:8

[608] Mishne Torah, Hilchot Melachim, 11:3

But the Messiah will be no ordinary man. He will be a king, but will be honored more than other kings.[609] He will have more wisdom than did King Solomon.[610] He will not only be a guide for all men, but he will somehow get all the nations of the world to serve God together.[611] "If a king arises from the House of David who meditates on the Torah, occupies himself with the commandments as did his ancestor King David, observes the commandments of the written and oral law, prevails upon all Israel to walk in the way of the Torah and to follow its direction, and fights the wars of God, it may be assumed that he is the Messiah. If he does these things and is fully successful, rebuilds the Third Temple on its location, and gathers the exiled Jews, he is beyond a doubt the Messiah. But if he is not successful, or if he is killed, he is not the Messiah."[612] This is clearly a significant departure from both the Christian and the Islamic eschatological beliefs.

"In the generation in which the son of David comes, the face of the people will be as that of a dog," says the Gemara.[613] Holocaust era rabbi Elchonon Wasserman believed that pre-Messianic Era leaders would be those who had turned away from God: "The teachers, the guides, the writers, the party leaders, it is they who prevent the radiance of the Torah from penetrating the darkness of men's hearts. They have their own Torah . . . Through their medium of a new Torah and new precepts they cause darkness to rule in the mind and in the heart . . . these are the leaders whom the prophet foresaw for our generation."[614]

Are we now living in the run up to the Messianic Era? The reestablishment of the Jewish homeland is believed by many to be testimony to that fact. The prayer for the modern State of Israel calls it, "the first flowering of our redemption," and calls for "peace, wise and principled leadership, a strong and safeguarded defense force, the ingathering of exiles, a faithfulness and closeness to God and His commandments, the

---

[609] Rambam's commentary, Mishne Sanhedrin Chapter 10

[610] Mishne Torah, Hilchot Tshuva, 9:2

[611] Mishne Torah, Hilchot Melachim, 11:4

[612] Mishne Torah, Hilchot Melachim, 11:4

[613] Talmud Bavli, Sotah, Chelek

[614] Ikvessa D'Meshicha

coming of the Messiah and the final redemption, and the acceptance of God's revelation throughout the world."[615]

Yet the messianic dream has been severely tempered by the diminishing of Jewish exceptionalism. It is at best acknowledged as a religious precept, symbolic but without practical application. Berel Wein sums it up nicely: "In a strange and almost irrational manner, the Jewish people favored being under foreign rule and its 'protection' over true national independence and reliance upon their own abilities and God's protective hand, so to speak... Much of the ambivalence that is present today in the Jewish world regarding the State of Israel stems from this fear of independence and longing to belong to a foreign nation that will somehow alleviate our problems and make us less special."[616]

5. THE LAND

There is only one religion whose scripture promises a specific piece of property to its constituency.

In his *Darko Shel Mikra*, Martin Buber reflects on this uniqueness by noting the ancient practice of bringing bikkurim (first fruits) to the Temple in Jerusalem: "Gifts offered to the gods from the first of the harvest are a familiar phenomenon of all cultures... as are prayers... thanking the gods for the blessing of the land... and asking them to ensure that the land remains fertile. But of all these types of prayers in the world, I know of only one in which the worshipper praises God for having given him a LAND... The speaker does not say merely 'I have come to the land,' but rather he states that he 'declares' to God that he has come to the land. The significance of this is as follows: I testify and identify myself as a person who has come to the land... He does this because he has to say, 'Not only the nation of Israel, but also this man who stands here has come to the land. I, the individual, identify myself as someone who has come to the land, and from time to time, when I bring the first of its fruits, I recognize this fact anew and declare it anew...' Every farmer in every generation of Israel thanks God when he brings his bikkurim for THE land to which He (God) brought HIM."

---

[615] The Koren Yom Haatzmaut Mahzor, p. 387

[616] Berel Wein, Jewish Destiny Blog, Shlach, June 13, 2015

The chosen promise includes redemption and does so in the context of a particular place – the land of Israel.

Scripture designates Abraham as the progenitor of the people that God subsequently identifies as His personal possession. And the highlight of that "chosen" narrative is the promise of a homeland. In Genesis 12 Abraham is instructed to leave his land and "go to the land that I will show you." Three chapters later (Genesis 15:18) God makes a covenant with Abraham, giving that land to his descendants. The past tense of the word "given" is used in the text, as if to say that this action had already been taken -- at creation according to many biblical commentators. By the time the Jews are ready to enter the Promised Land, they are told, "The eyes of God are always on it, from the beginning of the year to the end of the year." (Deuteronomy 11:12) "It is a very, very good land." (Numbers 14:7) "It is a blessed land." (Deuteronomy 33:13) The actual geographic boundaries of the land are given in Genesis 15, Exodus 23, Numbers 34, and Ezekiel 47. In Deuteronomy 1:8 the land is promised again to the Israelites.

In Genesis 28:4 Isaac promises Jacob the land that God gave to Abraham, his father. God delivers a personal guarantee to that effect to Jacob in Genesis 28:13-15. God promises that He will never break this covenant with the Jews (Leviticus 26:44-45 and Deuteronomy 30:3-5). The people that God had set aside from all others were provided a land on which to dwell. Yet, to satisfy the doubts of those who did not believe in his God, Abraham purchased property in the land in the presence of gentile witnesses (Genesis 49:30). God promised the land seven times to Abraham, once to Isaac and three times to Jacob. The Torah is the story of the journey, both spiritual and historic, that takes this people to their promised land.

Jerusalem is mentioned 349 times in the Bible and by the name *Tzion* (Zion) an additional 108 times. Yet it is not mentioned by name in the first five of those 24 books at all. Allusion is made to it in Genesis 4:18 when Abraham meets king Malchizedik of Shalem. According to tradition the binding of Isaac also takes place there. Mt. Moriah/Jerusalem is associated with other seminal events involving the biblical forefathers, including Jacob's dream (Genesis 28:10-22) and Isaac in prayer before meeting his wife-to-be, Rebecca (Genesis 24:63-67).

David was directed to the spot that became the Temple Mount to build an altar to God. In the process he chose to purchase the property (2 Samuel, 24:18-25 and 1 Chronicles 21:18-27). Psalm 137, which is said at every Jewish wedding ceremony, equates forgetting Jerusalem with forgetting one's own right hand.

According to scripture, the land of Israel was promised to the Jews in perpetuity. God directed the Jews to identify a particular place in the land where He would allow His name to rest – a place for the Holy Temple. The command to do so occurs often in Deuteronomy and in six separate verses of the 31 verses of Chapter 12 alone. Jerusalem was chosen by the Jews to be made holy by their presence and activities there as commanded in scripture.

The land promised to the Jews is a collective treasure. And the rejection of Zionism was seen by many as a betrayal of that legacy. Said Rabbi Abraham Isaak Kook, "We do not need to eat the fruits and be sated with the goodness (of an easy and tranquil life among the nations. Were this so) one could then make the blasphemous claim, 'Have we not found tranquility and goodness among the Gentiles, so what need have we for the land of Israel?' Heaven forbid! We do not pour out our hearts and wait in anticipation all our days for an illusory material tranquility. Rather our hope is to dwell in the presence of God there, the place designated for His service and for the observance of His Torah."[617]

The emancipation of the Jews in 18th century Europe was an individual emancipation, permitting Jewish participation as individuals only. Moses Hess argued passionately in his *Rome and Jerusalem: The Last National Question* that such an emancipation was illusory. Only national emancipation was possible for the Jews. He called the denial of Jewish nationalism traitorous. Rabbi Kook recognized that the weakness of the exceptionalist claim to the land of Israel was largely psychological. He called it an exilic introversion that caused an over-spiritualization of the biblical mandate. The study of text had replaced the actual performance of religious acts. As retold by Jonathan Sacks in his *One People?*, Kook believed that the secular settlement of the land was to be admired as an effort to arrest the continuing atrophy of Judaism.

---

[617] *One People?* quoting Eleh Divrei Haberit, pp. 154-5

The Elements of Jewish Exceptionalism

"Eretz Israel (the land of Israel) is not something apart from the soul of the Jewish people; it is no mere national possession, serving as a means of unifying our people and buttressing its material, or even its spiritual, survival. Eretz Israel is part of the very essence of our nationhood; it is bound organically to its very life and inner being. Human reason, even at its most sublime, cannot begin to understand the unique holiness of Eretz Israel; it cannot stir the depths of love for the land that are dormant within our people."[618]

For more than a century, Jewish immigrants to the former British Mandate have joined the indigenous Jewish community in developing a largely forgotten land for the benefit of themselves and their progeny. The resulting environmental and infrastructural improvements eclipse any comparable global enterprise on record. External investments in both capital and manpower have been repaid many times over by the technological, methodological and intellectual fruit of this national restoration effort.

Simultaneously billions upon billions of dollars invested over seven decades in the Arab portion of the former British Mandate have returned almost nothing. The standard of living has barely risen while development contributions are directed toward a reality-defying effort to remove Israel and its Jews from the region.

The reality of modern Israel is breathtaking. Modern highways, public transit, water recycling, education and research fuel a burgeoning standard of living. Neighborhoods spring up overnight. Attacks on individual and collective security are met with new and tested responses. The repatriation of East Jerusalem after 19 years of neglect has created an accessible magnet for the inheritors of the world's biblical legacy. The Jewish appreciation of history is in full flower as the antiquities of dozens of ancient cultures are discovered, restored and made accessible to all. In contrast, the rulers of ancient Gaza and the Temple Mount systematically destroy the vestiges of a 4000 year old Canaanite village and the remains of the Temples of Solomon and Herod.

The fact that the Jewish right to the land of Israel is routinely disputed or negated by international political and religious alliances is a modern, historical anomaly. Even if the nations of the world choose to

---

[618] *The Zionist Idea*, "The Land of Israel," Abraham Isaac Kook, p. 419

ignore the religious legacy of the Jewish people, they are still left to face the fact that Jews have had an unbroken presence in the land for 3000 years. Overwhelming historical and archeological evidence for the claim of Jewish title to the land is ignored, as is two centuries of entrepreneurial and communal settlement and development, independent of any colonizing power. If the fact of that development is still denied, the granting of political sovereignty to the Jews by the international community early in the twentieth century and again in the middle of the century is the legal recognition of its status as a Jewish state. Finally, even if that political legitimacy is withdrawn, the territory which comprises the Jewish homeland was captured in a defensive war. No nation on earth has ever been able to offer so convincing an argument for its legitimacy.

The effort to denigrate the Jewish claim to the land of Israel includes the work and writings of Jewish academics and historians. The Zionist narrative is erroneous, they claim, because there has been regular ethnic discontinuity in the makeup of the nation in the course of more than two millennia. The Hasmonean dynasty was known to force those it conquered to convert to Judaism. The Khazar kingdom of middle Europe in the sixth through 11th centuries was a nation of converts from both paganism and the surrounding Muslim and Christian empires. Today's Jews are more likely to trace their DNA to those converts than to their Palestinian forerunners. Early Zionists were not above suppressing these historic facts in order to preserve the notion that contemporary Jews were ethnically linked to those of the First Temple period. But this objection does nothing to weaken the exceptionalist argument and its attachment to the land. Critics fail to recognize that Jewish nationhood is not based on ethnicity, but on divine selection – a designation that comprises defined elements of an eternal legacy. Those attached to this divine designation are part of a historically unique people, a family that has absorbed adherents over the ages, each of whom is a legitimate owner of that identity. [619]

---

[619] See *The Invention of the Jewish People*, by Shlomo Sand, Verso, London, 2009. He argues that history denies both Jewish identification as a people indigenous to the land as well as their right to identification as a unique ethnicity.

Anti-Semitism and "The Jewish Problem"

"It is painful, and frightening, to watch European nations struggle to contain the contradictions that can come with liberalism . . . yes, there are excruciating reasons for these distinctions. But it is one thing for opprobrium to be unevenly distributed. Can any minority truly be secure when rights are unevenly distributed as well?"[620] The editor of *The Atlantic*, James Benner, generalizes the suffering of all minorities in distinguishing today's anti-Semitism from that preceding the Second World War. As a social critic, he's missed the salient historic and religious lesson: neither the protests of the media nor those of "enlightened" European leaders can disguise an embedded discomfort with the Jews and the embodiment of their exceptionalism, the State of Israel. Today and every day is 1933. A momentary redirection of the political and economic winds can quickly reopen old wounds and set in motion the vitriol, condemnation and physical abuse so common to the Jewish experience.

One need not be a philosopher or a psychologist to understand anti-Semitism. One need only be aware of religious history. Admittedly, this is a bitter pill to swallow in 21st century egalitarian, democratic society.

General displeasure with the Jews has evolved into a tripartite campaign to eliminate Jewish secular and religious influence. The traditional anti-Semitism of the religious right continues. Even the "conciliatory" writings and pronouncements of Catholic leaders retain the declaration that the Church has replaced the Jews as the Chosen of God. The Muslim world stands opposed to the existence of the State of Israel as a colonial imposition on its hegemonic embrace of the Middle East and institutionalizes the Qu'ranic demonization of the Jews. The liberal left in North America, Europe and the South Pacific accuses the Jew of directing the disenfranchisement and oppression of minorities in what is known as the "intersectional" grouping of oppressed populations. Jews are complicit in the maltreatment of Blacks in the United States and the denial of equal rights to women and LGBT individuals all over

---

[620] "To Stay or to Go," *The Atlantic*, James Bennet, Editor, April, 2015, p. 8

the world. This new anti-Semitism is both hypocritical and devoid of historical fact or perspective, yet its roots are ancient.

Much to the displeasure of the sociological and philosophical secular pluralists, an accurate understanding of anti-Semitism requires the acceptance of the biblical notion of creation and selection. In foregoing this principle, the scientific and philosophical intelligentsia conjures reasoned explanations for the foundations of Jew hatred. Yet, they can do the same for no other nation or ethnicity.

Understanding comes only from axiomatic acceptance: The Jews were assigned a singular role that conferred upon them responsibility to learn and live the contents of a divinely dictated or inspired rule book. This is the essence of "chosenness." Without this designation, Jews would, in fact, be no different from any other nation or ethnic group. Without recognizing this distinction the nations of the world would be absent the motivation to persecute, subdue or destroy them. Lurking somewhere in the recesses of human DNA is an acknowledgement of and aversion to this divine mandate.

The creations of Christianity and Islam are direct challenges to the chosen Jewish designation. These alternative ideologies embody rejection of the divine covenant of the Jews and the Jewish claim to an exclusive divine relationship and status under that agreement. Both Christianity and Islam are replacement theologies. Each claims to supersede Judaism in the eyes of God.

The imprint of more than 3000 years of resentment and hatred toward Jews stemming from their biblically chosen status pervades all modern societies. It has been educationally, sociologically and subliminally imbedded in the flesh of successive generations of Christians and Muslims. The continued existence of the Jews in the face of damning and contrary religious doctrine is a slap in the face of competing theologies.

Because the scriptural Jewish narrative cannot be embraced by philosophers, scientists and academics, only manifest symptoms of anti-Semitism can be described and addressed. Who would be willing to accept self-abnegation by acceding to the notion that Christianity and Islam were in origin specifically contrived to replace Judaism? Faith in religious pluralism is the compromise offered to ensure that all religions are understood to be equally valid (which is a statement that all are equally invalid). It is from that starting point that commentators begin their analysis of the "Jewish Problem."

Sociologists argue that in the absence of the Jews another people would surely be found to take its place as a social, economic or racial scapegoat. Yet Jews are singled-out because of their biblical franchise. Without the development of early polytheism and later Christianity and Islam, there is no Jewish Problem. Acceptance of the unique Jewish role in creation is necessarily anathema to those who worship at the altar of religious pluralism.

The modern attempt to disassociate Zionism or the State of Israel from Judaism is a transparently weak attempt to disguise garden-variety anti-Semitism as opposition to a national movement. The world has been incredibly uncomfortable with the creation of the modern state of Israel and its regret has been palpable. American Secretary of State George Marshall told President Truman that he would resign if the U.S. gave its approval to the UN mandate, and the State Department has been trying to undo the damage done by that act ever since.[621]

President Barak Obama, in his groundbreaking 2009 address to the Arab nations at Cairo University declared that the creation of the State of Israel was recompense for Jewish suffering during the Holocaust. It was a European, colonial creation that displaced an indigenous people with a foreign one. In a single sentence, Jewish biblical and historical nationhood was erased in favor of the religious and contemporary Muslim narrative. Eight years later, outgoing Secretary of State, John Kerry doubled down on that belief, announcing that Israel could not be both a Jewish and democratic nation – effectively declaring that no legitimacy could ever be granted a Jewish state. His words conveyed the international understanding that there is no Jewish "nation" or "people" as the Bible would have it. As most Jews themselves believe, Judaism is nothing more

---

[621] In 1943, Rabbi Meir Bar-Ilan (Berlin) told Senator Robert Wagner of New York, "If horses were being slaughtered as are the Jews of Poland, there would by now be a loud demand for organized action against such cruelty to animals. Somehow, when it concerns Jews, everybody remains silent, including the intellectuals and humanitarians of free and enlightened America." Two years later, General George Patton diverted US troops to rescue 150 prized Lipizzaner dancing horses which were caught between Allied and Axis forces along the German-Czech border.

than a religion – a stateless belief system or cultural inheritance. The private understanding is that were this not the case, Jews would be under constant suspicion, representing a fifth column, abusing the benevolence of their host countries while awaiting their opportunity to aid an enemy.

The daily deligitimization of Israel and denigration of the Jews takes place in the most respected world forums. No other member nation of the United Nations has had its right to exist questioned. No other nation stands publically threatened with extinction on a daily basis by its neighbors. No other nation must defend the right to its own defense. No other nation is expected to not only warn its enemy of a response to an attack on its citizens, but also give the enemy sufficient time to escape injury. No other country must justify its "disproportionate response" to an existential threat to its own citizens or its existence. No other country is expected to sacrifice the safety of its own soldiers to avoid collateral damage to its enemies. No other country must daily be accused of participation in atrocities around the world in which it could have no conceivable interest. No other country is routinely accused of human rights violations by those who are themselves the most egregious offenders. No other country is regularly designated in opinion polls as the greatest threat to world peace.

And no other people is routinely targeted for abuse by citizens of whatever country they call home. No other people after being attacked must suffer the insistence of the leaders of the free world that such attacks are not specifically aimed at them. No other people is routinely demonized for the economic and social shortcomings of every country in the world.

After several years of planning, Canada opened its national Holocaust memorial in its capitol, Ottawa. Yet neither its signage or promotion mentioned the word "Jew." Even the unique design of the building, that of a star of David, was described only as a star of universality. The building, web site and promotion were scrubbed of all references to the Jews.

There is no solution to the problem of anti-Semitism. History teaches that neither the behavior of the Jews nor toward the Jews is relevant -- the Jews will not disappear. Neither will the "genetic" disposition of the gentile world change just because logic dictates that it must. Instead, Jews will continue to exist in a netherworld of uncertainty until an historic paradigm shift forces the ancient and divine designation on them once again.

VIII

THE FUTURE OF JEWISH EXCEPTIONALISM

The post-World War II stature of European Jewry has been permanently diminished. The State of Israel is politically, socially and journalistically demonized. In the eyes of the world Israel is the face of every Jew -- the actions and fate of one are those of the other. The effort to disentangle the two is debilitating for the Jew, but is at the same time disingenuous and hypocritical. The disparate and fractious efforts to exercise authority within the Jewish community are counterproductive and diminish the divine mandate. The lesson of Jewish history remains unlearned: unity trumps virtue.

Tractate Yevamot[622] of the Babylonian Talmud contains a lengthy discussion of prohibited sexual relationships and the biblical requirements of *chalitza* and *yibum* (the process of extricating oneself from or participating in levirate marriage). Beit Hillel and Beit Shammai (the two major schools of legal exegesis in Mishnaic times) disagreed in both interpretation and practice. In the context of this debate, Reish Lakish (a sage of the Mishna) offered that the *Book of Esther* is read on different days of the month of Adar (the month of Purim), specific to locations and circumstances. He then provided as explanation the words *Lo titgodedu* (Deuteronomy 14:1) -- a prohibition in the Torah ordinarily associated with cutting or bruising the body in the style of pagans.

---

[622] Talmud Bavli, Yevamot 13b

To Reish Lakish this phrase referred to the prohibition of the formation of separate sects or groups. He wanted to make it clear that one should not separate himself from the majority in pursuing an alternate practice, either in law or in custom. So, early in this discussion the reader gets the sense that it should be obvious to all that there is a method followed by the majority that is in opposition to many minority methods. And, because the majority rules in such rabbinic disputes, all others are expected to fall in line. However, in this case, nothing could be further from the truth!

The Gemara then offers examples that demonstrate stark differences in practice among the sages. Rabbi Eliezer is shown to have cut wood on the Sabbath for the forging of iron. Rabbi Yossi of Galil ate fowl with milk. Rabbi Akiva proscribed preparations for a circumcision on the eve of the Sabbath. Yet each respected the other's practice. In fact, the Gemara goes on to say that when others visited the places where these "unorthodox" behaviors prevailed, they were obligated to conduct themselves as did the locals. Even though they disagreed frequently -- in this case over permitted "levirate" marriages -- Beit Hillel and Beit Shammai allowed their followers to intermarry and they showed mutual respect for the other school's legal and customary positions.

"Just remember this rule: The sin of (disunity and) dissension is worse than the sin of idolatry, as we find by Achav and Shaul (The peace and harmony of Achav's generation caused its victory in war despite their idolatry) . . . 'The Holy One Blessed be He pardoned the sin of idolatry three times, but He did not forgive the sin of discord.'"[623] It will take more than good intentions to recapture the Jewish biblical mandate. It will take leadership and compromise. Will there be a communal understanding that those Jews whose practices differ from one's own have an equally valid place at the table?

It seems that the external pressure that has forced a convergence of the fate of the Jewish Diaspora and that of the state represents a very positive development. The land of Israel is a critical component of Jewish exceptionalism. Its reclamation as the Jewish homeland speaks to one of the characteristics of an exceptional people: Act rather than react. The reestablishment of the state is not a reaction to persecution so

---

[623] *Eim HaBanim Simeicha*, chapter 4 citing Shnei Luchot Habrit, Shaar HaOtiot 1:45a

much as it is an awakening of a hibernating spirit. The development of that reclamation project includes the restoration of the Hebrew language, another of the five components of biblical Jewish exceptionalism. While the modern incarnation of Hebrew is missing some of the divine genius of its biblical predecessor, it lays the groundwork for the resuscitation of a unique and proprietary body of knowledge.

Lip service is paid by the most observant Jews to the divinity of the Pentateuch. But practice does not always follow. Biblical narratives are more often than not portrayed as morality plays -- models of religious or moral behavior contrived to teach lessons rather than to describe the relationship of the Jews to the rest of the world. The creation and development of a Jewish diaspora guaranteed the withdrawal of scripture as a heritage with anything but the most figurative meaning. Similarly, the messianic ideal was scandalized by false messiahs and was spiritualized in the claims of Christianity and Islam. Finally, the likelihood of either modern intellectuals or religious leaders accepting the confluence of science and Judaism – specifically the Kabbalistic descriptions of creation and nature – is almost unimaginable. That would require scrapping two competing worldviews, one supporting institutionalized religion and the other promoting a very "bitter" humanism.

Nevertheless, as we see with the establishment of the state of Israel and the rebirth of the Hebrew language, the resurrection of Jewish exceptionalism has begun. To accelerate the process, a new educational emphasis must be applied. "We may have a Jewish learning, but as for a living Jewish knowledge, knowledge of Judaism for living -- if this were ours, then schools would be the first living proofs of it. But where are they, these first and most important proofs of Jewish knowledge?"[624] Samson Rafael Hirsch elaborates: "For the field of Jewish learning is not a lonely field, isolated from nature, history or real life. On the contrary--it invites its disciples to the contemplation of heaven and earth, to the survey of historical events and of the physical, spiritual, moral and social life of man in all its aspects, and encourages closeness of observation and exactness of knowledge."[625] The palaver and bromides of an impotent religious leadership encourage more focused faith and

---

[624] *Judaism Eternal*, Vol. II, p.157

[625] Ibid, p. 173

impassioned entreaties of God on behalf of the mother who mourns her murdered son or the son who mourns his murdered mother. This passive approach ignores the biblical mandate for unity, action and leadership.

In 1632, Renee Descartes was prepared to publish findings in which he concluded that all bodies travel through space in a straight line at a uniform speed. Half a century later Newton made this principle part of his description of the laws of nature.

Descartes suppressed his book for fear that he would share the fate of Galileo, who within a year was condemned by the Inquisition for insisting that the earth actually moves. Twelve years later, when he finally did publish his book, Descartes couched his conclusions in the language of "relative" motion, an inconsistent conclusion that Newton later ridiculed, not understanding Descartes' desire for self-preservation. "If the Inquisition had not condemned Galileo, Descartes would never have argued for the relativity of motion. But for the inconsistency of his system, Newton would not have made an issue out of absolute space and time. He would not have devised the bucket argument, Mach might never have had his novel idea, and Einstein would not have been inspired to his greatest creation. Had the Inquisition condemned Galileo a few months later, Descartes would have published his ideas in their original form – and general relativity might never have been found."[626]

Edward Gibbon, the 18th century historian and author of *The History of the Decline and Fall of the Roman Empire*, wrote that the study of history is "little more than the register of the crimes, follies and misfortunes of mankind."[627] The divine designation of Jewish chosenness/exceptionalism mandated that this small people would never be subject to the vagaries of such history. Instead, the historical paradigms of Judaism manifest the free will granted with creation and at the same time testify to the diminution and disappearance of the exceptional status granted them. The scriptural paradigm was dictated by God. Subsequent salves were human responses to historical crises including the emergence of Christianity and Islam. Rabbinic paradigms sought to

---

[626] *The End of Time*, p. 67

[627] *Commentary Magazine*, "The Best of Scribblers," Joseph Epstein, 9/2015, p. 50

guarantee the survival of adaptive tradition and did so, but at the cost of the scriptural, chosen designation. A paralyzing suppression of theological and communal creativity was invoked – and then incrementally legislated into existence. Over time, pressure built within the community until, like a pot of boiling water, it released an orgy of alternative thought and practice. Rabbinic paradigms were transformed, first by the Enlightenment and today, by the emergence of a generation of Torah-educated women and men who are politically and economically astute, equipped with a strong secular and scientific education, and who reject the contrived paradigm that emerged from the Holocaust.

Just as the suppression of the ideas of Descartes led to the development of lasting innovation in natural science, so too might centuries of suppression of free Jewish thought result in the embrace of divinely promised Jewish exceptionalism.

POSTSCRIPT

A single reality is described by the seemingly independent fields of religion, history, science, philosophy and sociology. These ostensibly contradictory disciplines have sometimes individually and often in concert subverted scriptural intent. It is not an overstatement to claim that this interruption has derailed and delayed the fate of mankind, often causing and elongating needless misery and suffering.

Unlike Christianity or Islam, Judaism is not a religion dependent on faith. It is a biblical imperative imposed on a single nation. Faith or its absence is secondary. The imperative is primary.

While millions of Jews lived in Babylonia twenty five hundred years ago, only 40,000 chose to return to Jerusalem once permitted to reestablish their national home. Eight centuries earlier, only 20% of the Jews of Egypt were willing to follow Moses out of slavery. Social normalization in the diaspora has comforted the Jewish people. Social acceptance has been a collective goal of the Jewish nation, both within and outside of the Holy Land since the destruction of the Second Temple. Yet the scriptural narrative offers no goal of ordinary and equal status for the Jew. Instead, this smallest of nations is to be the catalyst that propels the engine of realization and redemption. Jewish abandonment of the biblical chosen role derailed the scriptural plan more than two millennia ago. Circumstances that compel its revival are now stirring.

In the past century alone, four of the earlier described elements of Jewish exceptionalism have been revived:

- *Jews have returned to their biblical homeland.* The land has been reclaimed and a strong Jewish intellectual and economic identity has been asserted. A young nation is dealing with both

the internal and external pressures that shape its conduct, its spirit and its boundaries.
- *The Hebrew language has been reborn as a living tongue.* Hebrew is now both a conversational and literary tool for the majority of the world's Jews. It remains for the original biblical form to be mined for its extensive information content.
- *Frequent revelations point to the equivalence of the oral, Kabbalistic tradition and the science of sub-atomic physics.* Experts will explore that correspondence, shedding new light on what have been thought to be purely technical, material discoveries.
- *Archaeological finds bring the biblical narrative to life and confirm the details of a divine history formerly dismissed as fiction.* Ancient timelines come into conformance with biblical timelines allowing for convergence around the scriptural narrative.

With the realization of each scriptural, exceptional component, a path is being forged for the dissemination of the Messianic Idea. We know only that there is a biblical culmination to existence and that the Jews will be the catalyst for its implementation.

Exploring the chosen designation has also opened the door to existential understandings:

- *Where do we come from?* We are collections of bits of information, introduced into a matterless void and then assembled into purposeful substance.
- *How did that happen?* Information was supplied and manipulated by a force from outside of what we recognize as reality. The existence of our reality is completely dependent on that outside force and its provision of the information that initiated, comprises and maintains our reality.
- *Why are we here?* The outside force that brought reality into existence did so in order to have something to act upon. We are made in the image of that outside force, so we must exercise our

free will to act upon things that we create (the information that we manipulate).
- o *Where do we go from here?* Since information is indestructible, the stuff from which we are made is eternal. All of what we are today will exist forever. What tangible or intangible form that takes is pure speculation, but everything that you and I have learned and experienced in life is accessible to our contemporaries and successors.
- o *What about Providence?* The closer that a person sticks to the plan put in place by that outside force, the more likely will be a positive outcome. It is a statistical matter that compels and directs the exercise of free will.
- o *Are there any guarantees?* There can be none. How can there be complete certainty in a world built on free will? Anything that can happen will happen – whether we are there when it does is an aspect of providence – we can influence that personal outcome by better processing the information around us in the context of the divine plan.
- o *A chosen people?* The rules that apply to the Jews differ from those that apply to all other people. Divine expectations are by scriptural definition different. Jews interact with creation differently from all other human beings. As a result, providence affects them differently.

There is a Creator, an "other," who is outside of the system and who supplied and manipulated the "substance" of creation -- information. Arguments against the likelihood of there being a Creator can be both logically and circumstantially defeated. Is scripture a divine product? How else can modern science be so accurately reflected in ancient texts? Is there a chosen people? Scripture says so and supports the claim with the unique content of a singular people's history and language.

It sounds "sacrilegious," but institutional science and religion have partnered to steer mankind away from his fate. Man was created in the image of God – gathering and utilizing information to further the perfection of mankind and his environment. That has been done, but to a less than optimal degree. The Jewish people are intended to lead humanity toward the conduct and construction of a divinely desirable environment. This environment must accommodate all who live together in a universe brought into being and managed to further the existence of man. History cannot be ignored or reset. Christianity and Islam, created to wrest the chosen role from the Jews, dominate the post-biblical landscape. Understanding the history and sociology of replacement and rejectionist movements is a prerequisite for carrying out the chosen role – a role that must now be promulgated in conjunction with its usurpers. By understanding that Judaism is not only a faith but a nationality, and by accepting its rejection by the other nations of the world, Jews can begin to husband the chosen role.

Nietzsche wrote in his *Antichrist*, "Anti-Semitism is the final consequence of Judaism." Jewish denial of the chosen role leading to acceptance of religious pluralism and striving for complete integration in Diaspora societies has validated Nietzsche's conclusion. Can a single, marginalized nation, surviving only at the pleasure of its landed hosts marshal the substance of creation in order to return mankind to the path of scriptural intent? The future depends on a tiny nation accepting the sacrifices required in following that treacherous but rewarding path.

Unholy Alliance: Sources

- *Ages in Chaos, Immanuel Velikovsky, Doubleday, NY, 1952*
- *A Hidden Revolution, Ellis Rivkin, Abingdon Press, Nashville, 1978*
- *A History of Israel, J. Bright, 1972, The Westminster Press and SCM Press Ltd.*
- *A History of the Bible, Fred Gladstone, Boston, Beacon Press, 1959*
- *A New Cosmogony, Edward Fredkin, www.leptonica.com*
- *A Simple Theory of Consciousness, Piero Scaruffi, www.scaruffi.com, 2001*
- *A Test of Time: The Bible from Myth to History, David M. Rohl, Arrow Publishing, 1994 (originally 1955)*
- *Ante-Nicene Fathers, Christian Literature Company, 1885*
- *Anti-Judaism: The Western Tradition, David Nirenberg, W.W. Norton & Co., 2013*
- *B'Or Ha'Torah, Shamir, the Israel Association of Religious Professionals, Professor Herman Branover, Editor-in-Chief*
- *Behavioral and Brain Sciences, Cambridge University Press, U.K.*
- *Bible History Daily*
- *Breaking the Spell, Daniel C. Dennett, Penguin Books, 2007*
- *Brett Watson, The Mathematics of Monkeys and Shakespeare, 12/13/95, ww.nutters.org/monkeys.html*
- *Cambridge Ancient History*
- *Catastrophism & Ancient History, C&AH Press, Toronto, Canada, 1978-1993*
- *Centuries of Darkness, Peter James, I.J. Thorpe, Rutgers University Press, 1991*

- *Challenge and Transformation: SecondTemple and Rabbinic Judaism*, Lawrence H. Schiffman, lawrenceschiffman.com
- *Changing the Immutable*, Marc B. Shapiro, Littman Library, Oxford, 2015.
- *Christianity in Talmud and Midrash*, R. Travers Herford, London, Williams & Norgate, 1903
- *Christianity Today*
- *Christianity: Essence, History, Future*, Hans Kung, Continuum, 1995
- *CNET*
- *Coincidences in the Bible and in Biblical Hebrew*, Haim Shore, iUniverse, Lincoln, NE, 2008
- *Commentary Magazine*, Commentary, Inc.
- *Crime: Criminals and Criminal Justice*, Nathan Cantor, University of Buffalo, 1932
- *Dark Riddle: Hegel, Nietzsche, and the Jews*, Yirmiyahu Yovel, Cambridge: Polity Press, 1998
- *Darwin Day in America*, John G. West, ISI Books, 2007
- *Deutsch-Franzosische Jahrucher*, Karl Marx, 1844
- *Did Muhammad Exist? An Inquiry into Islam's Obscure Origins*, Robert Spencer, ISI Books, 2012
- *Dreams of a Final Theory*, Steven Weinberg, Vintage Books, March, 1994
- *Duties of the Heart*, R. Bachya ibn Paquda, Feldheim Publishers, Jerusalem, 1962
- *Egypt Under the Pharaohs*, Heinrich Brugsch bey, 1891, Kessinger Publishing reprint, 200
- *Eim HaBanim Simeicha*, Yisachar Shlomo Teichtal, Translated by Moshe Lichtman, Kol Mevaser Publishers, Jerusalem, 2000
- *Eitz Chaim*, Rabbi Isaac Luria
- *El-Mas'udi's Historical Encyclopedia: The Meadows of Gold and Mines of Gems*, Mas'udi Ali-Abu'l-Hassan, reprinted Ulan Press, 2011
- *Emunot V'Deot, The Book of Beliefs & Opinions*, Saadia Gaon, Yale University Press, 1948

- *Essential Papers on Messianic Movements*, Marc Saperstein, Editor, New York University Press, New York, 1992
- *Evolution: the First Four Billion Years*, Michael Ruse (Editor), Harvard University Press, 2009
- *Ex nihilo Tech Journal*, Creation Revolution, Liberty Alliance
- *Fabric of Reality: The Science of Parallel Universes and its Implications*, David Deutsch, Viking Adult Publishing, 1997
- *Faith after the Holocaust*, Eliezer Berkovits, Ktav Publishing House, New York, 1973
- *Famines in Early History of Egypt and Syro-Palestine*, William Shea, PhD Dissertation, University of Michigan, 1976
- *Free Will*, Sam Harris, Free Press, 2012
- *Genesis and the Big Bang*, Gerald Schroeder, Bantam Books, 1991 reprint
- *God is Not Great: How Religion Poisons Everything*, Christopher Hitchens, Twelve, Hatchette Book Group, 2007
- *God, Freedom, and Evil*, Alvin Plantinga, Eerdmans, 1977
- *Hakirah: The Flatbush Journal of Jewish Law and Thought*, Hakirah Inc., New York
- *Hellenism and Christianity*, Gerald Friedlander, P. Vallentine & Sons, London, 1912
- *History of the Christian Religion to the Year Two Hundred*, Charles B. Waite, Book Tree, 1992
- *Horeb*, Samson Rafael Hirsch, Soncino Press, Seventh Edition, 2002
- *Ikvessa D'Meshicha*, (English Translation) Elchanan Wasserman
- *Immortality, Resurrection and the Age of the Universe: A Kabbalistic View*, Aryeh Kaplan, KTAV Publishing, NJ, 1993
- *In Crime: Criminals and Criminal Justice*, Nathan Cantor, University of Buffalo, 1932
- *Innerspace*, Aryeh Kaplan, Moznaim Publishers, 1991
- *Introduction to the History of Science*, George Sarton, Krieger Publishing Company, reprinted 1975, Volume II
- *Is the Bible True*, Jeffrey L. Sheler, Zondervan Publishing, 2000
- *Is There a God (revised edition)*, Richard Swinburne, Oxford University Press, 2010

- *Jerusalem Post*
- *Jesus the Messiah: A Survey of the Life of Christ*, Robert Stein, IVP Academic, 1996
- *Jewish Action Magazine, The Union of Orthodox Congregations*
- *Jewish Encyclopedia*, 1906 Edition
- *Jewish Thought*, Orthodox Union
- *Journal of Biosocial Science*, Cambridge University Press, U.K.
- *Journal of Chinese Medicine*
- *Journal of Economic History*, Economic History Association, Cambridge University Press
- *Journal of Scientific Exploration*
- *Judaism Eternal*, S.R. Hirsch, Vol. II, translated by Dr. I. Grunfeld, Soncino Publishers, 1959
- *Judaism: A Quarterly Journal of Jewish Life and Thought*, American Jewish Congress
- *Kabbalah*, Gershom Scholem, New American Library, NY, 1974
- *Kol Hatur*, Rabbi Hillel of Shklov
- *Kuzari*, Yehudah HaLevi, N. Daniel Korobkin Translation and Annotation, Jason Aronson Publishers, 1998
- *Language in the Americas*, J.H. Greenberg, Stanford CA, Stanford U Press, 1987. M. Ruhlen, editor
- *Maimonides Confrontation with Mysticism*, Menachem Kellner, Oxford University Press, 2006
- *Maimonides Introduction to the Mishnah*, Translated by Avraham Yaakov Finkel, Yeshivath Beth Moshe, 1993
- *Maimonides: Torah and the Philosophical Quest*, David Hartman, Jewish Publication Society, 1976
- *Mathematics and the Divine: A Historical Study*, T. Koetsier, editor, Luc Bergmans, Elsevier Science, 2005
- *Michtav Me-Eliyahu, Strive for Truth*, Rabbi Eliyahu E. Dessler, Feldheim Publishers, NY, 1978
- *Midbar Shur*, Jerusalem, 1997
- *Mind and Cosmos*, Thomas Nagel, Oxford University Press, 2012
- *Mind Magazine*, John Leslie Mackie, Evil and Omnipotence, 1955
- *Mindful Universe: Quantum Mechanics and the Participating Observer*, Henry P. Stapp, Springer, 2007

- *Moreh Nevuchim, Guide of the Perplexed,* Maimonides, Translated by Shlomo Pines, University of Chicago Press, 1963
- *Muhammad,* Michael Cook, Oxford University Press, 1983
- *My Philosophical Development,* Bertrand Russell, Routledge, NY, 1995
- *Myths, Models and Paradigms,* Ian G. Barbour, HarperCollins College Division, 1974
- *National Jewish Post & Observer*
- *National Public Radio*
- *Nature Magazine*
- *Nefesh HaChaim,* Rabbi Chaim Velozin
- *New Scientist*
- *Newsweek Magazine*
- *No Meek Messiah,* Michael Paulkovich, Spillix, LLC, Annapolis, MD, 2012
- *On the Origin of Species,* Charles Darwin, 1859, 150th Anniversary Edition, Signet, 2003
- *One People? Tradition, Modernity, and Jewish Unity,* Jonathan Sacks, Littman Library of Jewish Education, London, 1993
- *Orot,* A.Y. Kook, translation and notes by Rabbi Bezalel Naor, Aronson, 1993
- *Orthodox Union Weekly Newsletter*
- *Otzar HaChaim,* Isaac of Akko
- *Our Religions,* Arvind Sharma, Editor, HarperCollins Publishers, 1993
- *Permission to Believe,* Lawrence Kelemen, Targum/Feldheim, 1990
- *Persons: Human and Divine,* Oxford: Clarendon Press, 2007, "Idealism Vindicated," Robert M. Adams, editors Peter van Inwagen and Dean Zimmerman
- *Pharaohs and Kings,* David Rohl: A Biblical Quest, Crown Publishers, 1995
- *Philosophies of Judaism,* Julius Guttmann, Jewish Publication Society, 1964
- *Physics Today*

- *Plato's Timaeus, Stanford Encyclopedia of Philosophy, full text and translation at www.ellopos.net*
- *Rabbi Nachman's Stories, Aryeh Kaplan, Breslov Research Institute, 1983*
- *Rabbi Yechiel Poupko, "The Song of Chana: A review of the prayerful life of Chana in Sefer Shmuel, September, 2014.*
- *Rare Earth: Why Complex life is Uncommon in the Universe, Peter D. Ward and Donald Brownlee, Copernicus Books, 2000*
- *Revolutionary Antisemitism in Germany: From Kant to Wagner, Paul Lawrence Rose, Princeton University Press, 1980*
- *Scientific American Magazine, Nature Publishing Group, NY*
- *Sefer HaChinuch, The Book of Mitzvah Education, Feldheim Publishers, 1978*
- *Song of the Soul, Yechiel Bar-Lev, reprint edition 1994*
- *Sparks of the Hidden Light, Ateret Tiferet Institute, Moshe Schatz, 1998*
- *Sulam edition of the Zohar*
- *Surpassing Wonder: The Invention of the Bible and the Talmuds, Donald Harman Akenson, University of Chicago Press, 1998*
- *Synchronized Chronology: Rethinking Middle East Antiquity: a Simple Correction to Egyptian Chronology Resolves the Major Problems in Biblical and Greek Archaeology, Roger Henry, Algora Publishing, 2002*
- *Tanya, Rabbi Jacob Immanuel Schochet*
- *The Admonitions of an Egyptian Sage, Sir Alan Henderson Gardiner, Scholar's Choice, reprint 2015*
- *The Anguish of the Jews, Edward H. Flannery, Stimulus Books, Paulist Press, New York, 1985*
- *The Arabs in History, 4th revised addition, Bernard Lewis, Oxford University Press, 1966*
- *The Aryeh Kaplan Anthology, Mesorah Publications, New York, 1975*
- *The Astonishing Hypothesis: The Scientific Search for the Soul, Francis Crick, Simon & Schuster, 1994*
- *The Atlantic*

- *The Book of Jubilees, The Little Genesis, The Apocalypse of Moses, Joseph B. Lumpkin, 2006, Fifth Estate, AL*
- *The Collected Works of Samuel Taylor Coleridge: Table Talk, Edited by Kathleen Coburn & B Winer, Princeton University Press, 1991*
- *The Economist Magazine, London, U.K.*
- *The End of Faith: Religion, Terror, and the Future of Reason, Sam Harris, W.W. Norton, 2005*
- *The End of Time, Julian Barbour, Oxford University Press, 1999*
- *The Evolution of the Mother Tongue, NY, John Wiley & Sons, 1994.*
- *The Existence of God, Richard Swinburne, Oxford University Press, 1979*
- *The Exodus Problem and its Ramifications, Donovan Courville, Loma Linda, 1971*
- *The First 3 Minutes: A Modern View of the Origin of the Universe, Steven Weinberg, Basic Books, 1993*
- *The Foundations of Bible History, John Garstang, Kregel Publishing, 1978*
- *The God Delusion, Richard Dawkins, Mariner Books, Houghton Mifflin, New York, 2006*
- *The Grand Design, Stephen Hawking, Leonard Mlodinow, Bantam Books, 2010*
- *The Group Mind, William McDougall, Cambridge University, 1920, University of Michigan Library reprint*
- *The Holographic Universe, Michael Talbot, Harper Perennial, reprint 2011*
- *The Hundredth Monkey, Ken Keyes, Jr., Vision Books, 1982*
- *The Jews of Islam, Bernard Lewis, Princeton U Press, New Jersey, 1984*
- *The Koren Mahzor for Yom Haatzmaut and Yom Yerushalayim, Koren Publishers Jerusalem, 2015*
- *The Modern Jew Faces Eternal Problems, Aron Barth, Zionist Organization of Jerusalem, 1956*
- *The Moral Landscape, Sam Harris, Free Press, 2010*

- *The Moses Legacy: The Evidence of History,* Graham Phillips, Pan Books, 2003
- *The Muslim World,* Wiley, Vol 41, 1951
- *The Nature of Space and Time,* Stephen Hawking and Roger Penrose, The Isaac Newton Institute Series of Lectures, Princeton University Press, 1996
- *The New Atheism and Five Arguments for God,* William Lane Craig, reasonablefaith.org, 2010
- *The Old Kingdom from Abraham to Hezekiah – A Historical and Stratigraphical Revision,* Damien Mackey, 12/2002
- *The Order of Nature and Moral Luck: Maimonides on Divine Providence,* Steven Nadler, University of Wisconsin-Madison, 2011
- *The Origin of Speeches, Intelligent Design in Language,* Isaac E. Mozeson, Second Edition, Lightcatcher Books, 2011
- *The Physical Nature of Consciousness,* Evan Walker, John Benjamins Publishing Company, 2001
- *The Proof of God: The Debate that Shaped Modern Belief,* Larry Witham, Atlas Books, 2008
- *The Pyramid Builders of Ancient Egypt: A Modern Investigation of Pharaoh's Workforce,* Dr. A. Rosalie David, Routledge, revised, 1997
- *The Quantum Soul: A Scientific Hypothesis,* Stuart Hameroff, Deepak Chopra, Springer, 2011
- *The Sages,* Ephraim E. Urbach, The Magnes Press, 1975
- *The Structure of Scientific Revolutions,* Thomas Kuhn, University of Chicago Press, 1962, Third Edition, 1996
- *The Syro-Aramaic Reading of the Koran: A Contribution to the Decoding of the Language of the Koran,* Christoph Luxenberg, Prometheus Books, 2009
- *The Way of God: Derech Hashem,* Rabbi Moshe Chaim Luzzatto, Philipp Feldheim, 2009
- *The Western Canon: The Books and School of the Ages,* Harold Bloom, Riverhead Books, 1995
- *The Works of Josephus,* translated by William Whiston, Hendrickson Publishers, MA, 1987

- *The Zionist Idea*, Ed. Arthur Hertzberg, Atheneum, NY, 1972
- *Theology & Religion in Biblical Hebrew*, Stuart Weeks, Durham University
- *Time Magazine*
- *Torah Study Center*, Menachem Liebtag, Yeshivat Har Etzion, Israel
- *Tradition Magazine*
- *Treatise on the Gods*, H.L. Menken, 1930, reprinted Johns Hopkins University Press, 1997
- *U.S. News and World Report*
- *Union College Magazine*, Union College, Schenectady, NY
- *Universes*, John Leslie, Routledge, 2002
- *Wanderings: Chaim Potok's History of the Jews*, Chaim Potok, Knopf, New York, 1978
- *Why Aren't Jewish Women Circumcised? Gender and Covenant in Judaism*, Shaye J.D. Cohen, University of California Press, 2005
- *Windows into Old Testament History;* Long, Baker, Wenham, Editors, William B. Eerdmans Publishing, 2002
- *www.wired.com*
- *www.aish.com*
- *www.Algemeiner.com*
- *www.PBS.org*
- *www.PhysLink.com.*
- *www.rabbiwein.com*
- *www.reasons.org*
- *www.salon.com*
- *www.torahinmotion.org*
- *www.TorahWeb.org*